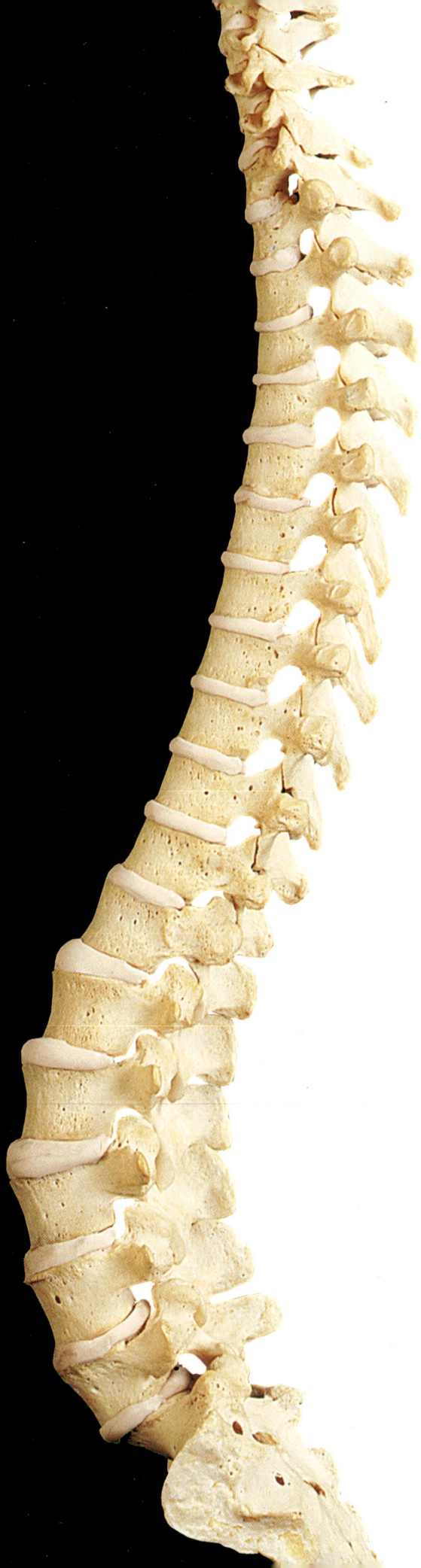

D0024800

A Brief Atlas of the

HUMAN BODY

Matt Hutchinson
Jon Mallatt
Elaine N. Marieb
Patricia Brady Wilhelm

Photographs by
Ralph T. Hutchings and Nina Zanetti

PEARSON
Benjamin Cummings

San Francisco Boston New York
Cape Town Hong Kong London Madrid Mexico City
Montreal Munich Paris Singapore Sydney Tokyo Toronto

Editor-in-Chief: Serina Beauparlant
Project Editor: Karoliina Tuovinen
Assistant Editor: Jessica Brunner
Managing Editor, Production: Wendy Earl
Production Manager: Janet Vail
Cover and Front Matter Designer: Yvo Riezebos Design

Photography Credits: Figures 1 – 71*: Ralph T. Hutchings
Plates 1 – 55: Nina Zanetti, Benjamin Cummings
Publishers, Pearson Education

*Except Figure 57 from *The Bassett Atlas of Human Anatomy* by Robert A. Chase, ©1989, The Benjamin/Cummings Publishing Company, Inc., and Figure 68a photographed by John Martinek, Kirkwood Community College.

ISBN 0-8053-7373-X
20 21 22 23 24 25 26 27 28 29 30 31 32 33 34 35—DOW—17 16 15 14 13

Copyright © 2007 Pearson Education, Inc., publishing as Benjamin Cummings, San Francisco, CA 94111. All rights reserved. Printed in the United States of America. This publication is protected by copyright and permission should be obtained from the publisher prior to any prohibited reproduction, storage in a retrieval system, or transmission in any form or by any means, electronic, mechanical, photocopying, recording, or likewise. For information regarding permission(s), write to: Pearson Education, Inc., Rights and Permissions Department.

Pearson Benjamin Cummings is a trademark of Pearson Education, Inc.
www.aw-bc.com

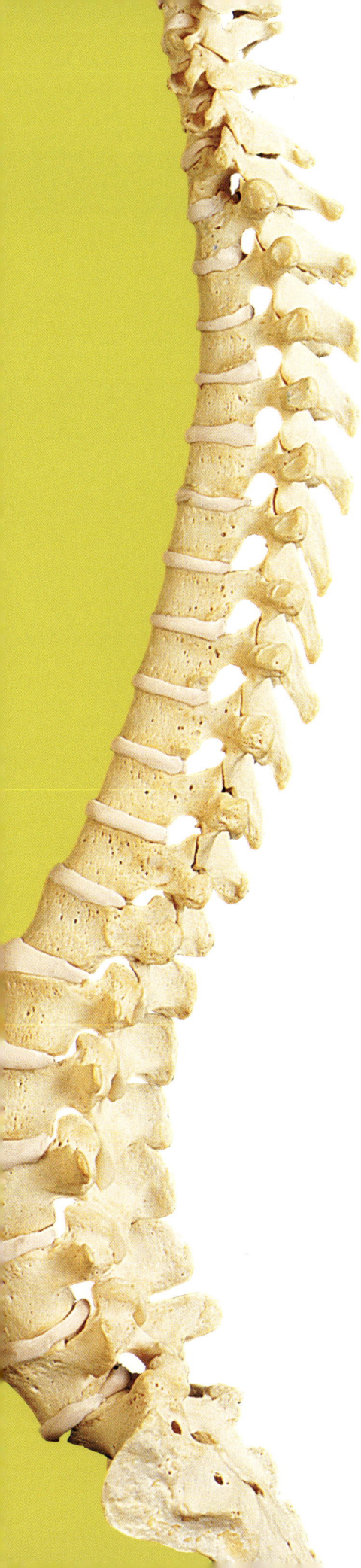

PREFACE

Building upon our original vision, the second edition of *A Brief Atlas of the Human Body* features 51 soft tissue images and 104 bone images providing a degree of clarity and scale that could never be achieved within a textbook alone. In addition, this edition includes a brand new section, histology of basic tissues and select organs, containing 55 outstanding histology slides photographed by Nina Zanetti of Siena College.

Elaine Marieb chose many of the soft tissue views. Matt Hutchinson, of Washington State University, took on the arduous task of labeling each structure. Jon Mallatt, co-author of the human anatomy text, scrutinized and approved the views, leaders, and labels. References to related atlas images herein can be found in the illustration figure legends of both the human anatomy and the human anatomy and physiology textbooks.

For this edition, Patricia Brady Wilhelm, co-author of the human anatomy text, reviewed each photograph for accuracy and chose several new soft tissue images from the collection of Mark Nielsen and Shawn Miller of the University of Utah. She also worked to choose all of the slides included in the histology portion of the Atlas to represent the most useful selection of histology plates for students in the classroom.

Ralph T. Hutchings, formerly with The Royal College of Surgeons of England, photographed each of the bone structures and many of the soft tissue images found in this book. His reputation as an anatomical photographer preceded him, and we certainly were not disappointed—the quality of his work is here for all to see. We are most grateful to him for lending his expertise to this project, and for his good humor and ready willingness to meet our demands. We are grateful to John Martinek of Kirkwood Community College, who contributed his excellent photograph of the internal surface of the stomach (Figure 69a).

The authors would like to thank the following instructors for their expertise and thoughtful feedback in reviewing the Atlas: Andy Beall, University of North Florida; Dennis Carnes, Imperial Valley College; Leslie Hendon, The University of Alabama at Birmingham; H. Rodney Holmes, Waubonsee Community College; Jeff Kent, Volunteer State Community College; Mark Robertson, Delta College; Laura Rosillo, IVY Tech State College; Justine Wilcox, University of North Florida; Joseph Yavornitzky, Baldwin Wallace College; Hillman Mann, Volunteer State Community College; Darrell Davies, Kalamazoo Valley Community College; Louise Russo, Villanova University; Tom Swensen, Ithaca College.

We are hopeful that the second edition of *A Brief Atlas of the Human Body* proves to be a relevant source to students and instructors. Benjamin Cummings would welcome your comments and suggestions, which may be sent to the following address:

Publisher
Applied Sciences
Benjamin Cummings
1301 Sansome Street
San Francisco, CA 94111

CONTENTS

Part I HISTOLOGY OF BASIC TISSUES AND SELECT ORGANS 1

BASIC TISSUES 3

Epithelial Tissues 3

Plate 1 Simple squamous epithelium, surface view 3
Plate 2 Simple squamous epithelium, section through a renal corpuscle in the renal cortex 3
Plate 3 Simple cuboidal epithelium, l.s. through renal medulla 3
Plate 4 Non-ciliated simple columnar epithelium from the small intestine—jejunum 4
Plate 5 Ciliated simple columnar epithelium from the oviduct 4
Plate 6 Pseudostratified columnar epithelium from the trachea 4
Plate 7 Stratified squamous epithelium from the mucosa of the esophagus 5
Plate 8 Stratified cuboidal epithelium from the duct of a sweat gland 5
Plate 9 Stratified columnar epithelium from a duct in the parotid gland 5
Plate 10 Transitional epithelium from the ureter 6

Connective Tissues—Connective Tissue Proper 6

Plate 11 Areolar connective tissue 6
Plate 12 Adipose tissue from the external ear 6
Plate 13 Reticular connective tissue, lymph node 7
Plate 14 Dense irregular connective tissue from the submucosa of the large intestine 7
Plate 15 Dense regular connective tissue, tendon 7
Plate 16 Elastic connective tissue from the aorta 8

Connective Tissues—Cartilage 8

Plate 17 Hyaline cartilage from the trachea 8
Plate 18 Elastic cartilage 8
Plate 19 Fibrocartilage, within a tendon 9

Connective Tissues—Bone 9

Plate 20 Compact bone 9
Plate 21 Spongy bone 9

Connective Tissues—Blood 10

Plate 22 Blood smear 10
Plate 23 Blood smear, neutrophils 10
Plate 24 Blood smear, eosinophil 10
Plate 25 Blood smear, basophil 11
Plate 26 Blood smear, lymphocytes 11
Plate 27 Blood smear, monocyte 11

Muscle Tissue—Skeletal Muscle 12

Plate 28 Skeletal muscle l.s. 12
Plate 29 Skeletal muscle c.s. 12
Plate 30 Neuromuscular junction (motor end plate) 12

Muscle Tissue—Cardiac Muscle 13

Plate 31 Cardiac muscle 13

Muscle Tissue—Smooth Muscle 13

Plate 32 Smooth muscle, from the uterus 13

Nervous Tissue 13

Plate 33 Neuronal cell body in central nervous system 13
Plate 34 Peripheral nerve, c.s. 14
Plate 35 Peripheral nerve, l.s. 14

Select Organs 15

Plate 36 Spinal cord c.s. through lumbar region 15
Plate 37 Thick skin showing epidermal and dermal regions 15
Plate 38 Muscular artery and vein 16
Plate 39 Spleen 16
Plate 40 Trachea 16
Plate 41 Lung 17
Plate 42 Gastroesophageal junction 17
Plate 43 Stomach, l.s. 18
Plate 44 Small intestine, c.s. through duodenum 18

Plate 45 Mucosal layer of small intestine, from the jejunum 19
Plate 46 Pancreas 19
Plate 47 Pig liver 20
Plate 48 Renal cortex 20
Plate 49 Renal medulla 20
Plate 50 Urinary bladder 21
Plate 51 Testes 21
Plate 52 Ovary 22
Plate 53 Uterus, secretory stage 22
Plate 54 Thyroid gland 23
Plate 55 Adrenal gland, section 23

Part II BONES OF THE HUMAN SKELETON 25

Figure 1 Skull, anterior view 27
Figure 2 Skull, right external view of lateral surface 28
Figure 3 Skull, internal view of left lateral aspect 29
Figure 4 Skull, external view of base 30
Figure 5 Skull, internal view of base 31
Figure 6 Occipital bone, inferior external view 32
Figure 7 Frontal bone 33
(a) anterior view 33
(b) inferior surface 33
Figure 8 Temporal bone 34
(a) right lateral surface 34
(b) right medial view 35
Figure 9 Sphenoid bone 36
(a) superior view 36
(b) posterior view 36
(c) anterior view 36
Figure 10 Ethmoid bone 37
(a) left lateral surface 37
(b) posterior view 37
(c) anterior view 37
Figure 11 Mandible 38
(a) right lateral view 38
(b) right medial view 38
Figure 12 Maxilla 39
(a) right lateral view 39
(b) right medial view 39
Figure 13 Palatine bone 40
(a) right lateral view 40
(b) right posterior view 40
Figure 14 Bony orbit 41
Figure 15 Nasal cavity, left lateral wall 42
Figure 16 Fetal skull 43
(a) anterior view 43
(b) lateral view 43
Figure 17 Articulated vertebral column 44
(a) right lateral view 44
(b) posterior view 45
Figure 18 Various views of vertebrae C_1 and C_2 46
(a) atlas, superior view 46
(b) atlas, inferior view 46
(c) axis, superior view 47
(d) axis, inferior view 47
(e) articulated atlas and axis, superior view 47
Figure 19 Cervical vertebrae 48
(a) right lateral view of articulated cervical vertebrae 48
(b) fifth (typical) cervical vertebra, superior view 48
(c) fifth (typical) cervical vertebra, posterior view 48
(d) fifth (typical) cervical vertebra, right lateral view 49
(e) vertebra prominens (C_7), superior view 49
Figure 20 Thoracic vertebrae 50
(a) articulated thoracic vertebrae, right lateral view 50
(b) seventh (typical) thoracic verterbra, superior view 50
(c) seventh (typical) thoracic vertebra, posterior view 50
(d) comparison of T_1, T_7, and T_{12} in right lateral views 51
Figure 21 Lumbar vertebrae 52
(a) articulated lumbar vertebrae and rib cage, right lateral view 52
(b) second lumbar vertebra, superior view 52
(c) second lumbar vertebra, posterior view 53
(d) second lumbar vertebra, right lateral view 53

Figure 22 **Sacrum and coccyx** **54**
(a) posterior view 54
(b) right lateral view 54
(c) anterior view 55

Figure 23 **Thoracic cage** **56**
(a) anterior view 56
(b) posterior view 57
(c) sternum, right lateral view 58
(d) sternum, anterior view 58
(e) typical rib, posterior 59
(f) articulated typical rib and vertebra, superior view (left) and lateral view (right) 59

Figure 24 **Scapula and clavicle** **60**
(a) right scapula, anterior view 60
(b) right scapula, posterior view 60
(c) right scapula, lateral aspect 61
(d) right clavicle, inferior view (top) and superior view (bottom) 61
(e) articulated right clavicle and scapula, superior view 61

Figure 25 **Right humerus** **62**
(a) anterior view 62
(b) posterior view 62
(c) proximal end, anterior view 63
(d) proximal end, posterior view 63
(e) distal end, anterior view 63

Figure 26 **Right ulna and radius** **64**
(a) articulated right ulna and radius, anterior view 64
(b) articulated right ulna and radius, posterior view 64
(a) articulated right humerus, ulna, and radius, anterior view 64
(d) articulated right humerus, ulna, and radius, posterior view 64
(e) right ulna, proximal end: anterior, posterior, medial, and lateral views 65

Figure 27 **Bones of the right hand** **66**
(a) lateral aspect 66
(b) dorsal aspect 66

Figure 28 **Bones of the male pelvis** **67**
(a) right hip bone, lateral view 67
(b) right hip bone, medial view 67
(c) articulated male pelvis, anterior view 68
(d) articulated male pelvis, posterior view 69

Figure 29 **Right femur** **70**
(a) anterior view 70
(b) posterior view 70
(c) proximal end, anterior view 71
(d) proximal end, posterior view 71
(e) proximal end, medial view 71
(f) distal end, anterior view 72
(g) distal end, posterior view 72
(h) articulated right femur and patella, inferior view with knee extended 73
(i) right patella, anterior surface 73
(j) articulated right femur and patella, inferior posterior view with knee flexed 74
(k) right patella, posterior surface 74

Figure 30 **Right tibia and fibula** **75**
(a) articulated right tibia and fibula, anterior view 75
(b) articulated right tibia and fibula, posterior view 75
(c) right tibia, proximal end, anterior view 76
(d) right tibia, proximal end, posterior view 76
(e) right tibia, proximal end, articular surface 77
(f) articulated right tibia and fibula, proximal end, posterior view 77
(g) articulated right tibia and fibula, distal end, posterior view 77
(h) right fibula, proximal end, anterior view 78
(i) right fibula, proximal end, posteromedial view 78
(j) right fibula, proximal end, medial view 78

Figure 31 **Bones of the right ankle and foot** **79**
(a) superior surface 79
(b) inferior (plantar) surface 79
(c) medial view 80
(d) lateral view 80
(e) right calcaneus, superior aspect 81
(f) right calcaneus, posterior aspect 81
(g) right talus, inferior view 81

Part III SOFT TISSUE OF THE HUMAN BODY 83

Figure 32 **Superficial muscles of the thorax, posterior view** **85**

Figure 33 **Superficial muscles of the thorax, anterior view** **85**

Figure 34 **Abdominal muscles** **86**

Figure 35 **Right shoulder from right, showing deltoid muscle and biceps** **87**

Figure 36 **Triceps of the left arm, posterior view** **87**

Figure 37 **Right forearm and wrist** **88**
(a) palmar surface 88
(b) dorsum surface 89

Figure 38 **Wrist and hand** **90**
(a) dorsum surface of the right hand and wrist 90
(b) palmer surface of the left hand and wrist 91

Figure 39 **Superficial muscles of the superior gluteal region** **92**

Figure 40 **Superficial muscles of the left lower thigh, anterior view** **93**

Figure 41 **Right upper thigh, posterior view** **93**

Figure 42 **Leg** **94**
(a) medial view 94
(b) lateral view 94

Figure 43 **Foot** **95**
(a) anterior view 95
(b) medial view 95

Figure 44 **Right lower face and upper neck** **96**

Figure 45 **Muscles, blood vessels, and nerves of neck, anterior view** **97**

Figure 46 **Sagittal section of the head** **98**

Figure 47 **Left nasal cavity, lateral wall** **99**

Figure 48 **Right cerebral hemisphere (arachnoid mater removed)** **100**

Figure 49 **Ventral view of the brain** **101**

Figure 50 **Midsagittal section of the brain** **102**

Figure 51 **Transverse section of the brain, superior view** **103**

Figure 52 **Brainstem and cervical region of the spinal cord, posterior view** **104**

Figure 53 **Cervical region of spinal cord, ventral view** **105**

Figure 54 **Vertebral column and spinal cord** **106**
(a) cervical and upper thoracic regions from the left 106
(b) lower thoracic and upper lumbar regions from the left 106

Figure 55 **Spinal cord and cauda equina, dorsal view of lower end** **107**

Figure 56 **Heart and associated structures in thorax** **108**

Figure 57 **Heart and pericardium, anterior view** **109**

Figure 58 **Coronal section of the ventricles, anterior view** **110**

Figure 59 **Heart, posterior view (blood vessels injected)** **111**

Figure 60 **Pulmonary, aortic, and mitral valves of the heart, superior view** **112**

Figure 61 **Fibrous framework of the heart (atria removed), posterior view, from the right** **113**

Figure 62 **Tongue and laryngeal inlet** **114**

Figure 63 **Diaphragm, superior view** **115**

Figure 64 **Upper abdomen** **116**
(a) upper abdominal viscera, anterior view 116
(b) transverse section of upper abdomen, from below, showing liver, stomach, and spleen *in situ* 117

Figure 65 **Liver, posteroinferior view** **118**

Figure 66 **Lower abdominal organs** **119**

Figure 67 **Small intestine and colon** **120**

Figure 68 **Vessels of the gastrointestinal organs** **121**

Figure 69 **Internal surfaces of the stomach and small intestine** **122**
(a) frontal section of the internal surface of the stomach 122
(b) small intestine, cut open to show plicae circulares 123

Figure 70 **Retroperitoneal abdominal structures** **124**

Figure 71 **Kidney, internal structure in frontal section** **125**

Figure 72 **Male pelvis, saggital section** **126**

Figure 73 **Sections through male reproductive structures** **127**
(a) section through right testis and epididymis, and the penis
(b) transverse section through penis

Figure 74 **Female pelvis, sagittal section (uterus points forward in this view)** **128**

Figure 75 **Female pelvic cavity showing the position of the uterus relative to other structures** **129**

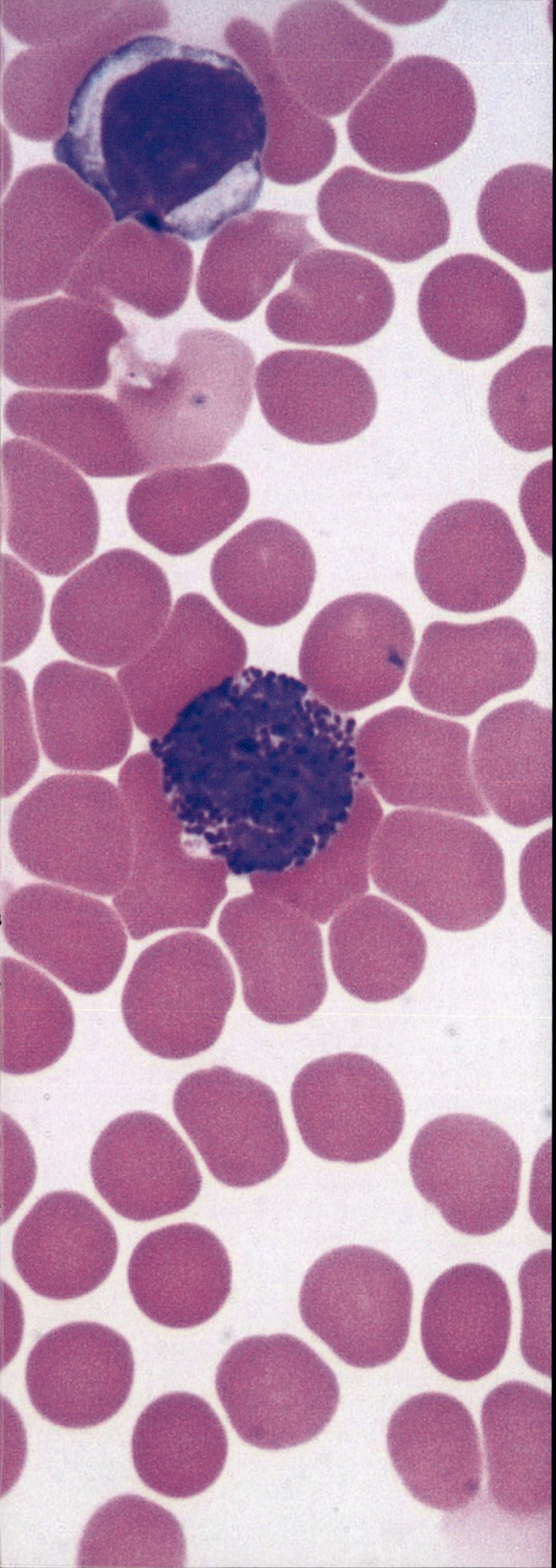

Part I

HISTOLOGY

BASIC TISSUES

EPITHELIAL TISSUES

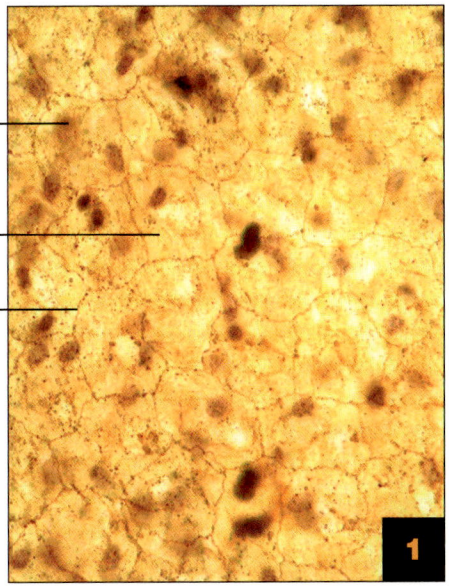

PLATE 1 Simple squamous epithelium, surface view. Silver stained mesothelium (360×)

DESCRIPTION: Single layer of flat cells with disc shaped central nuclei and little cytoplasm. In this surface view, the cells resemble fried eggs.

LOCATION: Form the kidney glomeruli and corpuscles; air sacs (alveoli) of the lungs; lining of the heart, blood vessels and lymphatic vessels; lining of the ventral body cavity (serosae).

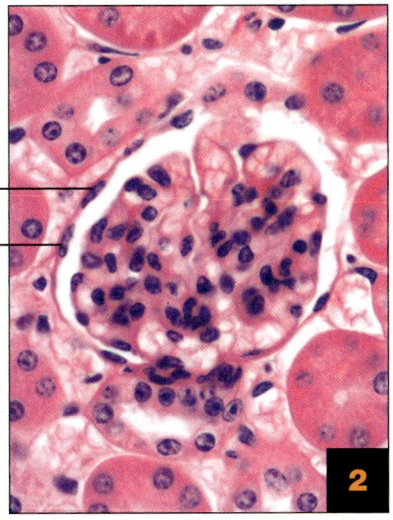

PLATE 2 Simple squamous epithelium, section through a renal corpuscle in the renal cortex (465×)

DESCRIPTION: The single layer of flat cells with disc shaped central nuclei and small amount of cytoplasm are apparent in the parietal layer of the renal capsule. In this section, the thinness of the squamous cells and the dark staining central nuclei are obvious.

LOCATION: As listed for Plate 1.

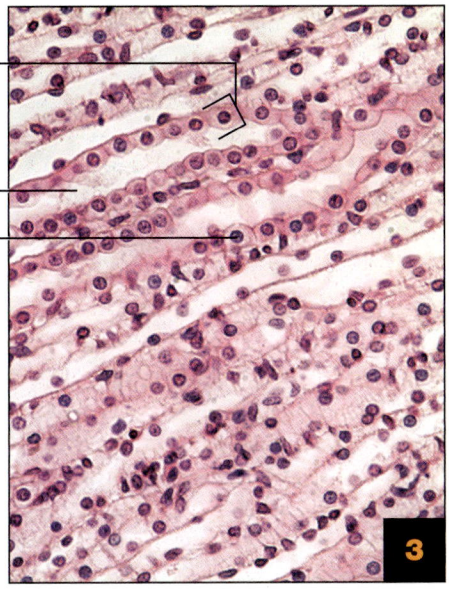

PLATE 3 Simple cuboidal epithelium, l.s. through renal medulla (350×)

DESCRIPTION: Single layer of cube-shaped cells with large round central nuclei. This section shows multiple rows of simple cuboidal epithelium forming the renal tubules.

LOCATION: Forms the kidney tubules and collecting ducts; the ducts and secretory portions of many glands; and the surface of the ovary.

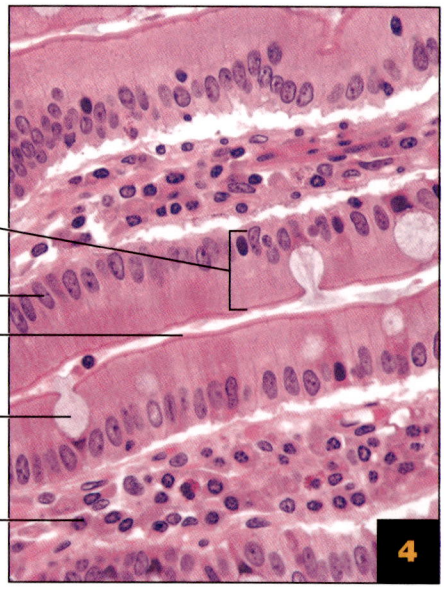

PLATE 4 Non-ciliated simple columnar epithelium from the small intestine—jejunum (360×)

DESCRIPTION: Single layer of column shaped cells with either a round or oval shaped nucleus. Unicellular glands (goblet cells) that secrete mucous are common in this tissue. Microvilli, extensions of the plasma membrane of the apical surface, are present in the small intestine.

LOCATION: Lines digestive tract from stomach to anal canal; the gallbladder; portions of uterus and uterine tubes; and the excretory ducts of some glands.

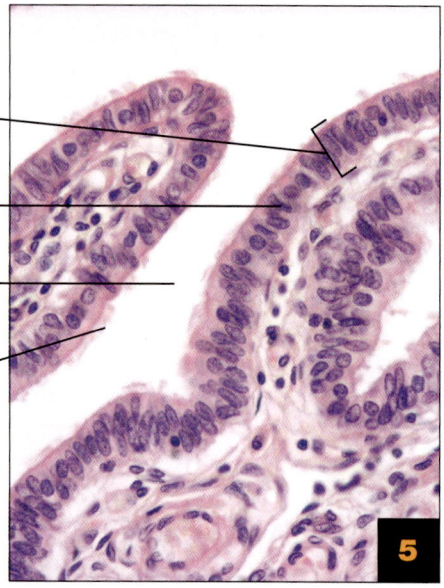

PLATE 5 Ciliated simple columnar epithelium from the oviduct (350×)

DESCRIPTION: Single layer of column shaped cells with either an oval or a round nucleus with cilia extending from the apical surface.

LOCATION: Lines the small bronchi, the uterine tubes, and portions of the uterus.

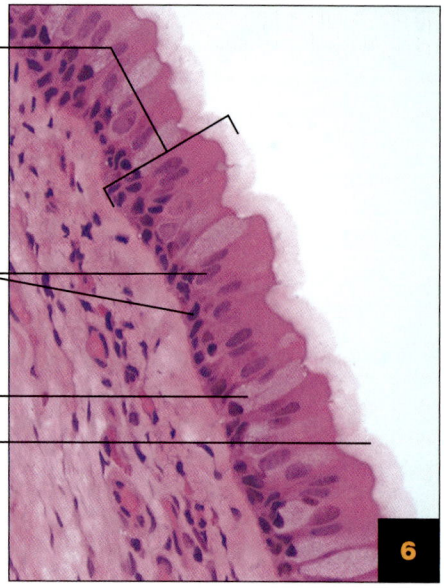

PLATE 6 Pseudostratified columnar epithelium from the trachea (335×)

DESCRIPTION: Single layer of cells of differing heights, some not reaching the apical surface. Nuclei located at different levels give the appearance of a multilayered (stratified) tissue. Mucous secreting goblet cells are common in this tissue. There are both ciliated and non-ciliated types. Goblet cells and cilia are seen in this specimen.

LOCATION: The non-ciliated type lines the sperm carrying ducts (duct of epididymis, ductus deferens, ejaculatory duct, and mid-portion of the male urethra); the ciliated type lines the trachea and most of the upper respiratory tract.

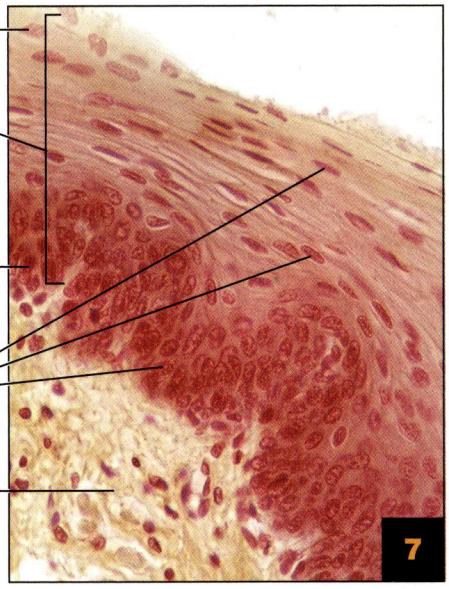

PLATE 7 Stratified squamous epithelium from the mucosa of the esophagus (350×)

DESCRIPTION: Distinguished by multiple layers of cells with nuclei distributed throughout. The basal cells are cuboidal or columnar in shape, are metabolically active and have a high rate of mitosis; the apical cells are flat (squamous). In the keratinized type the surface cells are filled with the protein keratin.

LOCATION: The non-keratinized type (seen in this specimen) lines the esophagus, mouth, vagina and anus. The keratinized type forms the epidermis of the skin.

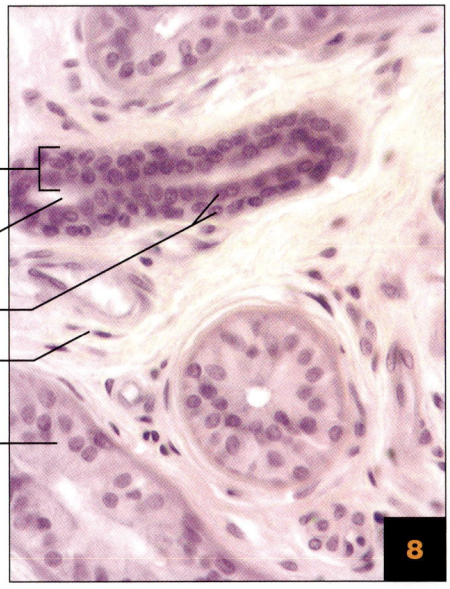

PLATE 8 Stratified cuboidal epithelium from the duct of a sweat gland (340×)

DESCRIPTION: Two layers of cells, identifiable by the two rows of round nuclei. The apical layer is composed of cube shaped cells.

LOCATION: Lines the largest ducts of sweat glands, the mammary glands, and the salivary glands.

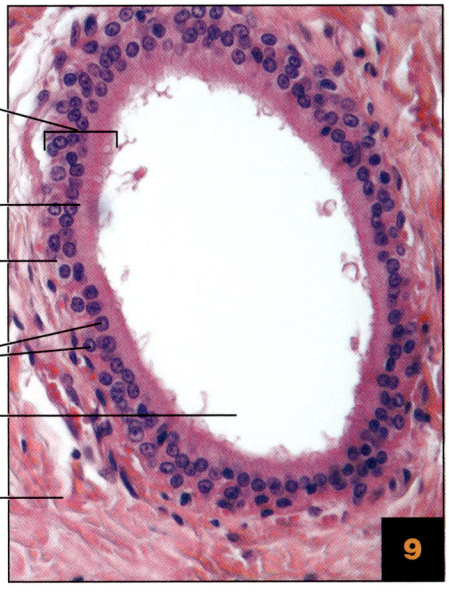

PLATE 9 Stratified columnar epithelium from a duct in the parotid gland (350×)

DESCRIPTION: Two or more layers of cells: the apical layer is composed of columnar shaped cells, the basal layer is usually cuboidal cells.

LOCATION: Lines the large ducts of some glands, and some portions of the male urethra.

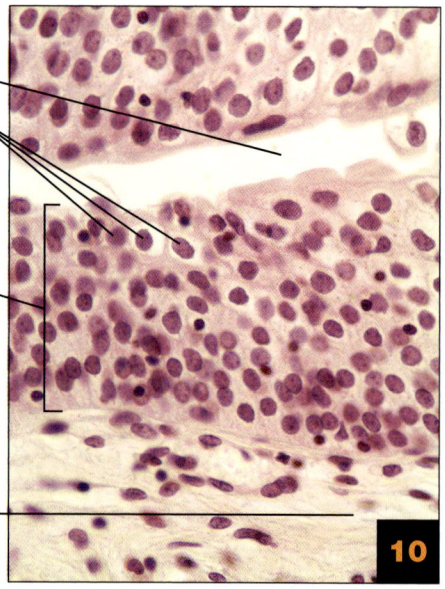

EPITHELIAL TISSUES, *continued*

PLATE 10 Transitional epithelium from the ureter (340×)

DESCRIPTION: As its name implies, this tissue changes shape. It is composed of multiple layers of cells. When in relaxed state, as shown here, the cells appear cuboidal in shape; when stretched, the cells appear squamous in shape.

LOCATION: Forms the lining of the urinary bladder, ureter, and the superior portion of the urethra.

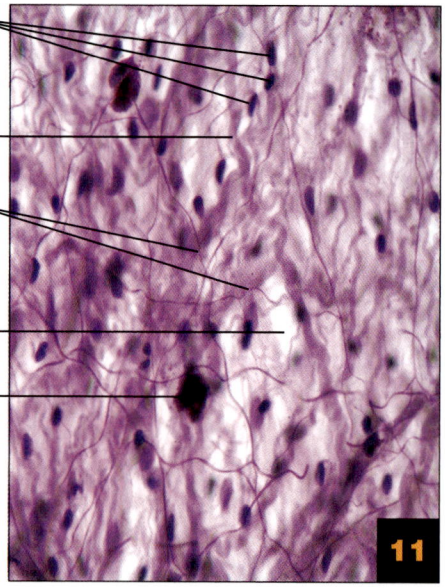

CONNECTIVE TISSUES—CONNECTIVE TISSUE PROPER

PLATE 11 Areolar connective tissue (350×)

DESCRIPTION: Matrix contains all three fiber types (collagen fibers, elastic fibers, and reticular fibers) within a gel-like ground substance. Fibroblasts, mast cells, macrophages and other white blood cells are found within this tissue.

LOCATION: Distributed throughout the body loosely binding adjacent structures: forms the lamina propria that underlies all epithelia in the body; forms the papillary layer of the dermis of the skin and contributes to the superficial fascia; surrounds blood vessels, nerves, muscles.

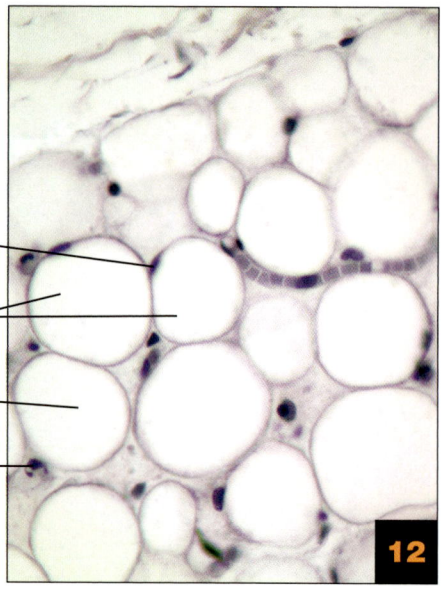

PLATE 12 Adipose tissue from the external ear (350×)

DESCRIPTION: Distinguished by the closely packed adipocytes (fat cells) within a sparse matrix. Each adipocyte is filled with a large fat droplet causing the nucleus to be pushed to the edge of the cell.

LOCATION: A ubiquitous tissue found throughout the body: the hypodermis of the skin; surrounding the kidneys, eyeballs, mammary glands, and many other body organs; within the abdomen; and within the fascial planes separating muscle layers.

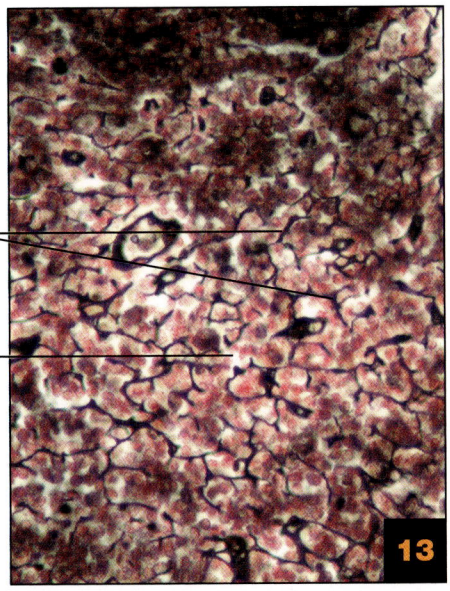

PLATE 13 Reticular connective tissue, lymph node (350×)

DESCRIPTION: Matrix is composed of reticular fibers loosely distributed within a gel-like ground substance. Cellular components are fibroblasts, lymphocytes, and other blood cells.

LOCATION: Forms the internal framework of many lymphoid organs: spleen, lymph nodes, bone marrow.

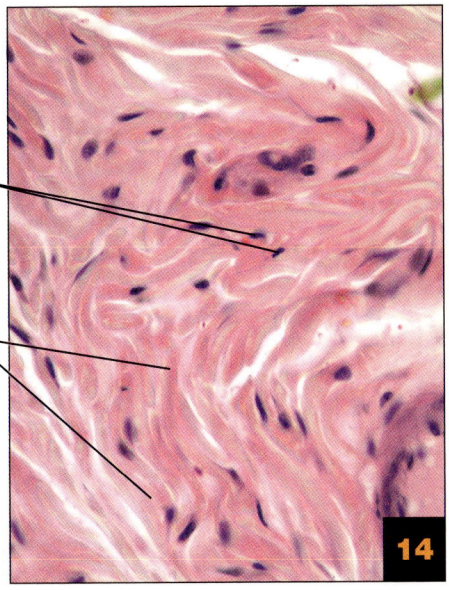

PLATE 14 Dense irregular connective tissue from the submucosa of the large intestine (350×)

DESCRIPTION: Distinguished by the irregular arrangement of fibers densely packed in multiple directions. Composed primarily of collagen fibers with some elastic fibers. Major cell type is the fibroblast.

LOCATION: Reticular layer of dermis of skin, submucosa of digestive tract, fibrous capsules of organs and joints.

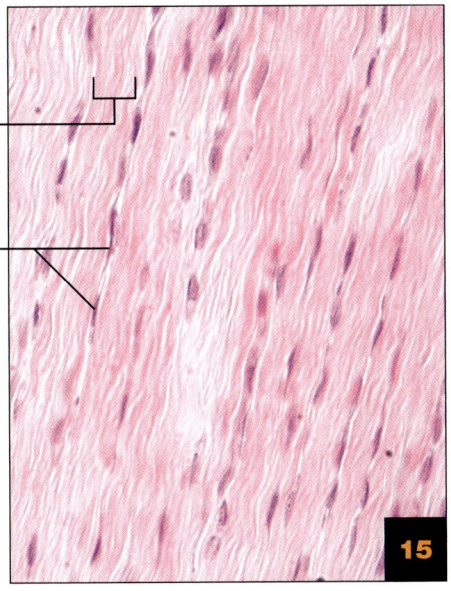

PLATE 15 Dense regular connective tissue, tendon (340×)

DESCRIPTION: Densely packed fibers, primarily collagen, arranged parallel to each other. The nuclei of the fibroblasts are also aligned in parallel. This is an important feature for differentiating this tissue from smooth muscle tissue.

LOCATION: Tendons, most ligaments, aponeuroses.

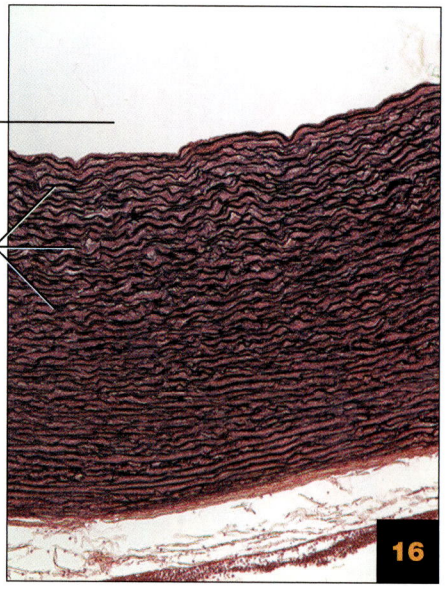

CONNECTIVE TISSUES—PROPER, *continued*

PLATE 16 Elastic connective tissue from the aorta (90×)

DESCRIPTION: Connective tissue with densely packed elastic fibers. Notice the wavy appearance of the dark staining elastic fibers. Here, these fibers are within the smooth muscle layer of the wall of the aorta.

LOCATION: Found within the body wall of arteries; distributed throughout the trachea and bronchial tree; located within some ligaments in the neck region.

CONNECTIVE TISSUES—CARTILAGE

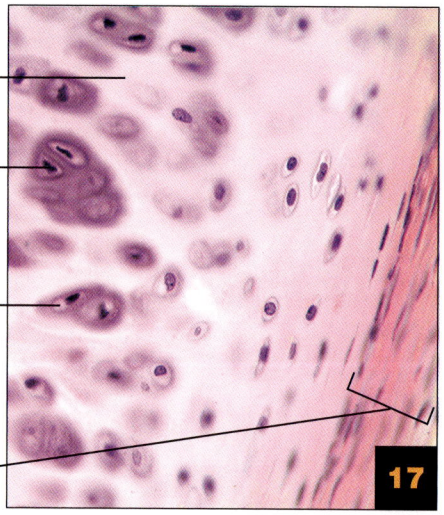

PLATE 17 Hyaline cartilage from the trachea (320×)

DESCRIPTION: Cartilage cells (chondrocytes) located within spaces (lacunae) in the tissue matrix. The matrix is a firm, gel-like ground substance embedded with collagen fibrils, which are not viewable via light microscopy. Vascularized perichondrium surrounds the cartilage nourishing the tissue and producing new tissue.

LOCATION: Forms most of the embryonic skeleton; covers the ends of long bones in joint cavities; forms the costal cartilages, the cartilages of the nose, trachea and bronchial tree, and most of the laryngeal cartilages.

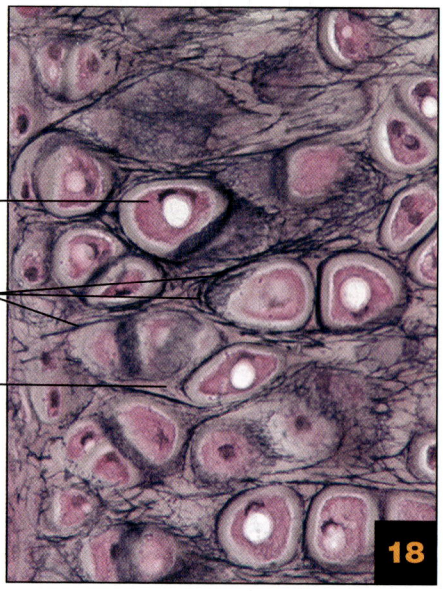

PLATE 18 Elastic cartilage (350×)

DESCRIPTION: As in hyaline cartilage, the chondrocytes sit in spaces (lacunae) within the tissue matrix. The matrix contains a firm gel-like ground substance and both collagen fibrils and elastic fibers. In this preparation the elastic fibers are the dark purple strands viewable in the matrix.

LOCATION: Epiglottis and the external ear.

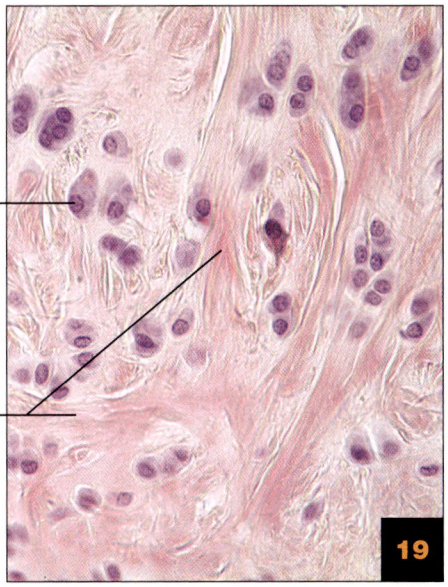

PLATE 19 Fibrocartilage, within a tendon (350×)

DESCRIPTION: The gelatinous matrix is densely packed with thick collagen fibers. Chondrocytes are located in lacunae, spaces in the matrix. This feature distinguishes this tissue from dense irregular connective tissue. This is the strongest type of cartilage.

LOCATION: Form the intervertebral discs; the pubic symphysis; the articular discs within joint cavities. Also located within some tendons, particularly where a tendon passes around a bony pulley.

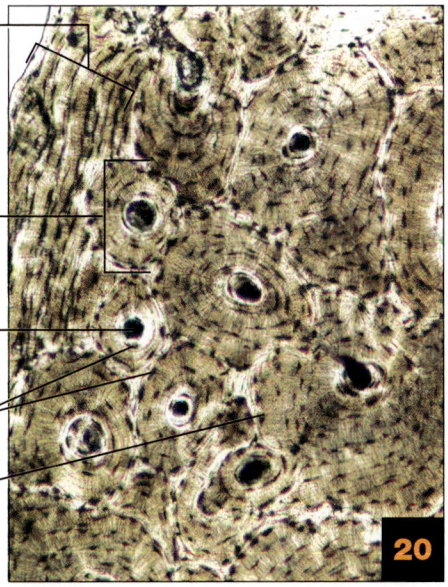

CONNECTIVE TISSUES—BONE

PLATE 20 Compact bone (90×)

DESCRIPTION: Tissue composed of a hard, calcified matrix containing many collagen fibers. This densely packed bone tissue is organized in lamellae (layers of bone tissue) and osteons (concentric rings of bone tissue). Blood vessels are located in the central canals; osteocytes lie in the lacunae; canaliculi, the thin dark lines, connect adjacent osteocytes.

LOCATION: Found in the shaft of long bones; the external portion of flat, short, and irregular shaped bones; and the external portion of the epiphyses.

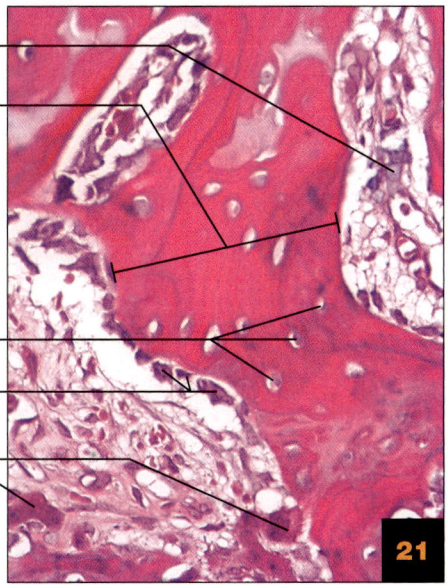

PLATE 21 Spongy bone (340×)

DESCRIPTION: Composed of the same materials (hard calcified ground substance and collagen fibers) as compact bone. Arranged into small beams (trabeculae) of bony tissue. Spaces between trabeculae are filled with red bone marrow. Bone arranged in lamellae and osteocytes are located in lacunae. Bone forming cells, osteoblasts, line the trabeculae. Bone destroying cells, osteoclasts, also present as spongy bone continually remodels.

LOCATION: Located in the internal regions of the epiphyses, as well in the internal portions of flat, short, and irregular shaped bone.

CONNECTIVE TISSUES—BLOOD

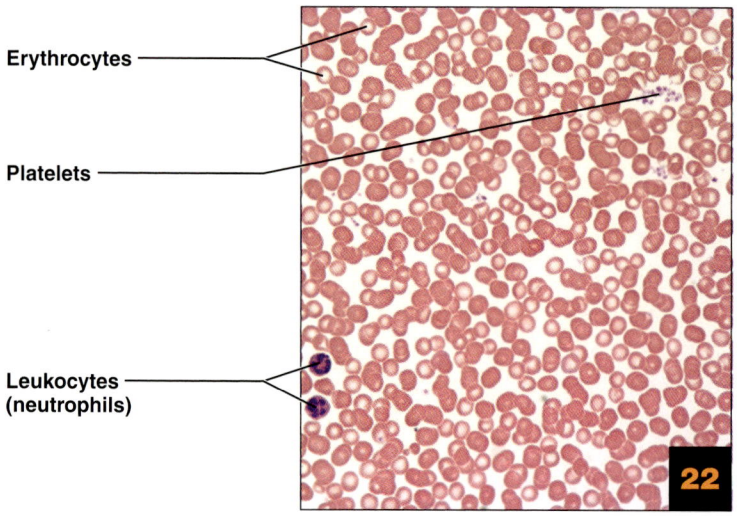

PLATE 22 Blood smear (270×)

DESCRIPTION: Erythrocytes (red blood cells), leukocytes (white blood cells), and platelets in a fluid matrix. Red blood cells, the small, red, disc-shaped cells shown here, are the most numerous formed element in blood.

LOCATION: Contained within blood vessels.

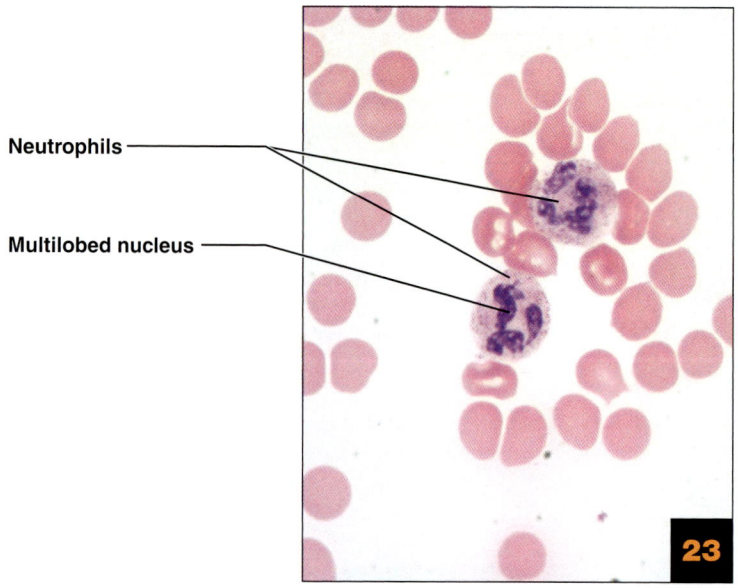

PLATE 23 Blood smear, neutrophils (895×)

DESCRIPTION: Granular leukocyte containing a multilobed nucleus. Cytoplasm appears light purple due to stain affinities. This is the most common leukocyte.

LOCATION: Originate and are stored in the red bone marrow. Travel in blood vessels. Enter loose connective tissues in response to infections. Respond to bacterial infections.

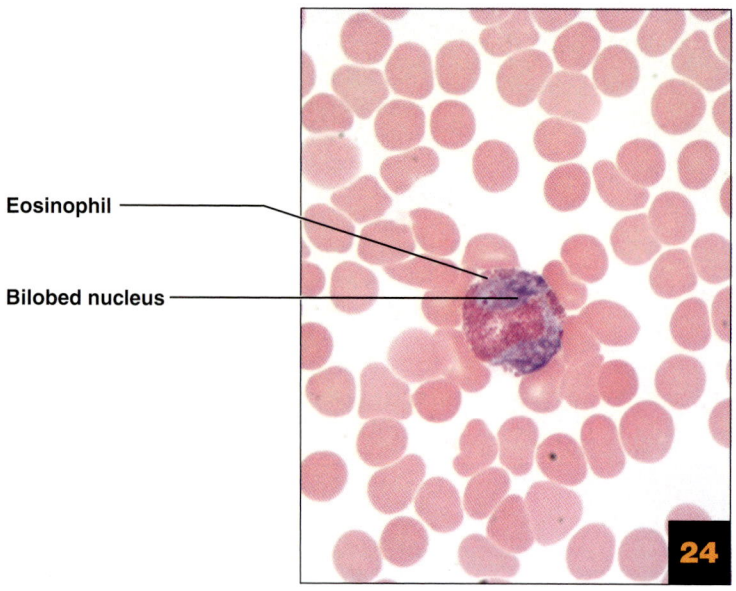

PLATE 24 Blood smear, eosinophil (885×)

DESCRIPTION: Granular leukocyte with a bilobed nucleus, cytoplasmic granules stain red. Relatively rare (1–4% of all leukocytes).

LOCATION: Originate and are stored in the red bone marrow. Travel in blood vessels. Enter loose connective tissues in response to infections. Respond to parasitic infection, and to end an allergic reaction.

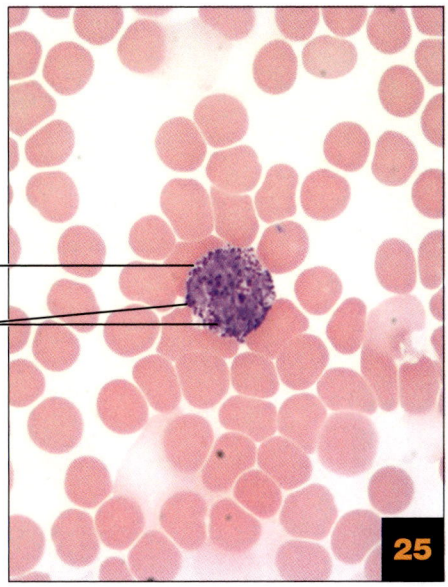

PLATE 25 Blood smear, basophil (840×)

DESCRIPTION: Granular leukocyte with a bilobed nucleus, cytoplasmic granules stain dark purple. Most rare (0–1% of all leukocytes).

LOCATION: Originate and are stored in the red bone marrow. Travel in blood vessels. Enter loose connective tissues in response to infections. Mediate inflammation by secreting histamines.

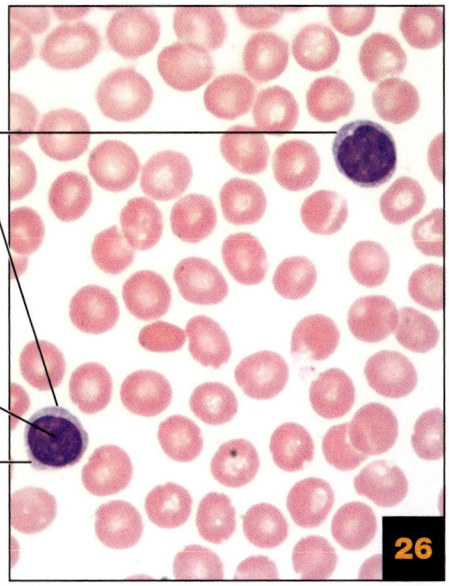

PLATE 26 Blood smear, lymphocytes (815×)

DESCRIPTION: Agranular leukocyte with a large circular nucleus that takes up most of the cell volume and stains purple, surrounded by a thin border of pale blue cytoplasm.

LOCATION: Originate in red bone marrow. Travel in blood vessels. Enter loose connective tissues and lymphoid tissue in response to infections.

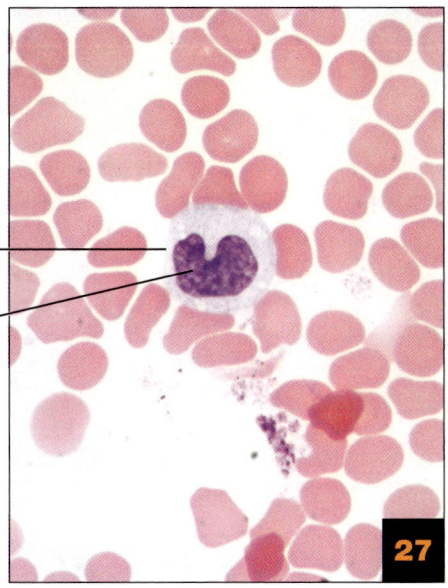

PLATE 27 Blood smear, monocyte (855×)

DESCRIPTION: Largest leukocyte; agranular with a kidney shaped nucleus that stains lighter than that of lymphocytes. Contains a larger amount of blue staining cytoplasm than in lymphocytes.

LOCATION: Originate and are stored in the red bone marrow. Travel in blood vessels. Enter loose connective tissues in response to infections, transform into macrophages to phagocytize foreign matter.

MUSCLE TISSUE—SKELETAL MUSCLE

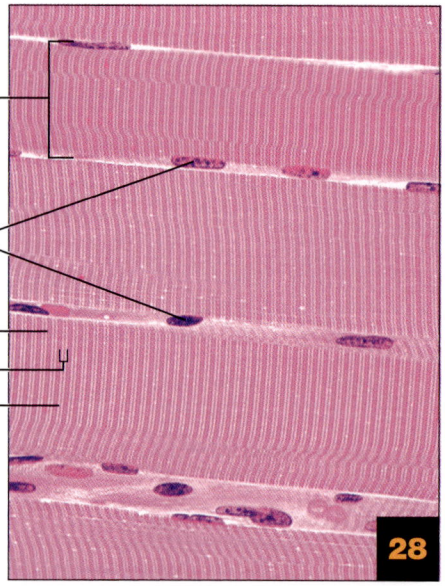

PLATE 28 Skeletal muscle l.s. (445×)

DESCRIPTION: Long cylindrical cells (fibers), multinucleated, obvious striations running perpendicular to fiber direction. Dark bands are called A bands, light bands are I bands. In this section you can also see the Z discs, the thin dark lines running through the middle of the I bands.

LOCATION: In skeletal muscles.

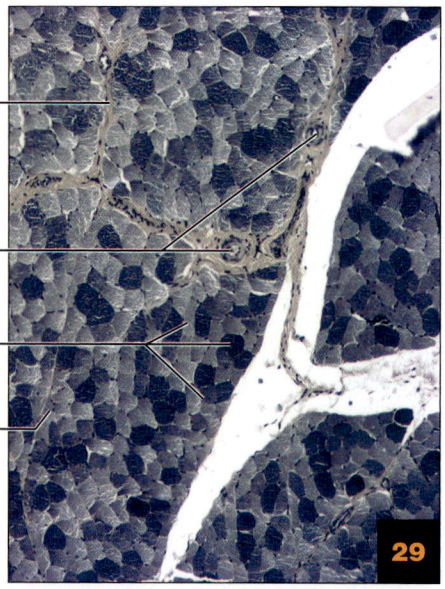

PLATE 29 Skeletal muscle c.s. (85×)

DESCRIPTION: Cross section through skeletal muscle showing muscle fibers and connective tissues. Nuclei are located at the periphery of each fiber and the connective tissue, perimysium, (stained gray) groups bundles of muscle fibers into fascicles. Notice the blood vessels running within the perimysium.

LOCATION: In skeletal muscles.

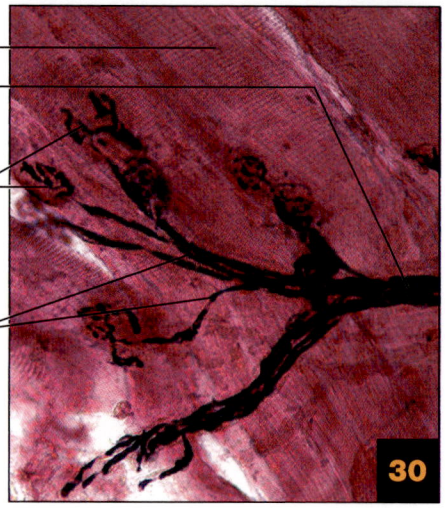

PLATE 30 Neuromuscular junction (motor end plate; 290×)

DESCRIPTION: Junction of a motor neuron with skeletal muscle fibers. The axon branches into multiple axon terminals that innervate muscle fibers.

LOCATION: Within skeletal muscles.

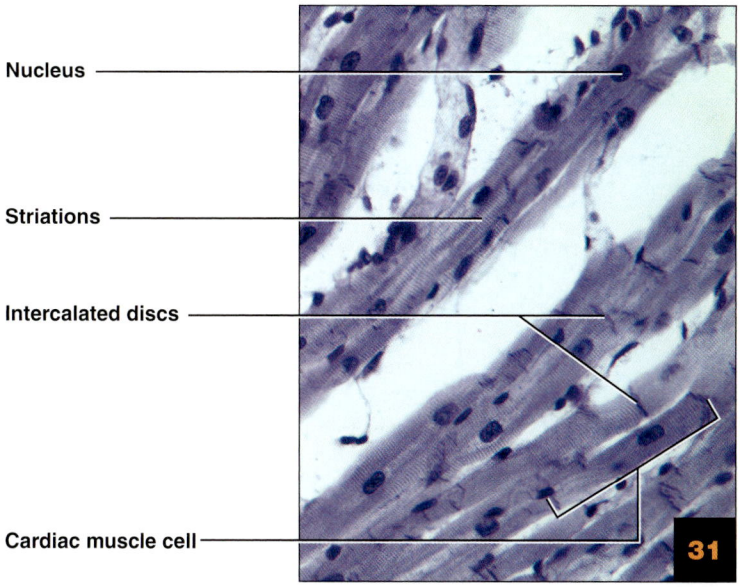

MUSCLE TISSUE—CARDIAC MUSCLE

PLATE 31 Cardiac muscle (365×)

DESCRIPTION: Striated muscle (although striations are often difficult to view at this magnification) composed of branching cells with one centrally located nucleus (occasionally two). Cells are joined by specialized cell junctions, intercalated discs.

LOCATION: Makes up the myocardium of the heart.

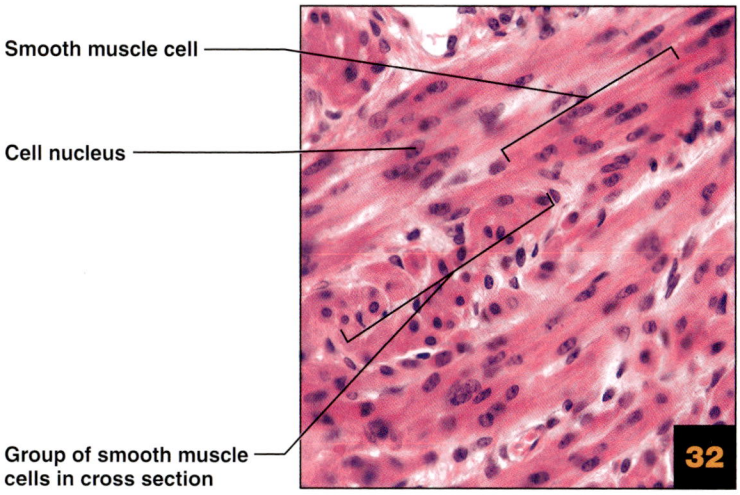

MUSCLE TISSUE—SMOOTH MUSCLE

PLATE 32 Smooth muscle, from the uterus (300×)

DESCRIPTION: Non-striated muscle tissue. Elongated, tapering cells with a single, central nucleus are closely packed together to form sheets. Distinguishable from dense regular connective tissue because nuclei are randomly distributed throughout.

LOCATION: Composes the muscular layer in the wall of the digestive tract, circulatory vessels, urinary, and reproductive organs. Also located in the respiratory tubes and inside the eye.

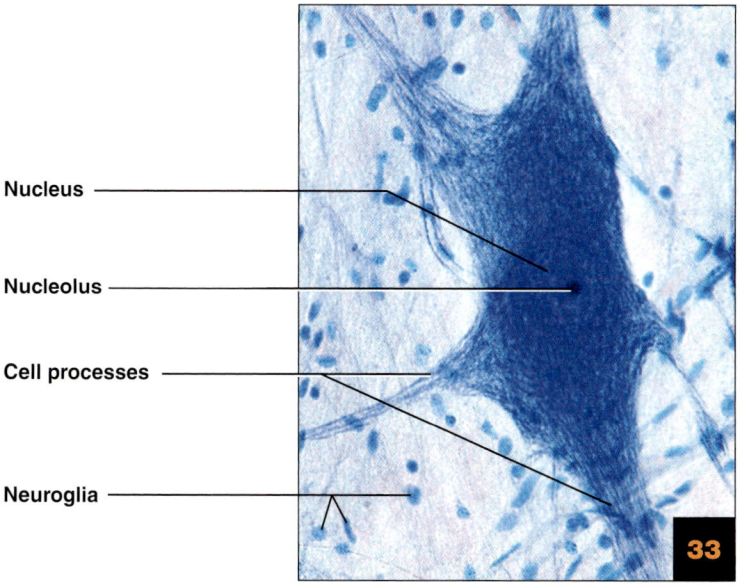

NERVOUS TISSUE

PLATE 33 Neuronal cell body in central nervous system (350×)

DESCRIPTION: The neuronal cell body contains the nucleus and cellular organelles. Extending from the cell body are cell processes that either receive or transmit signals. The dark staining fibers are neurofibrils, intermediate filaments that run throughout the neuron. This image also shows multiple neuroglia, supportive cells that aid neuronal function.

LOCATION: Found in the central nervous system in the brain and spinal cord; in the peripheral nervous system in peripheral ganglia.

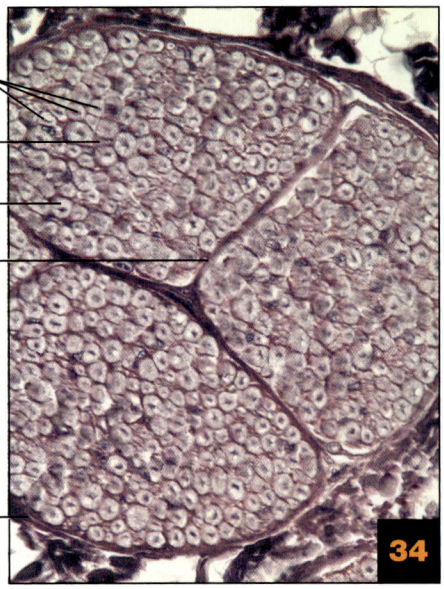

PLATE 34 Peripheral nerve, c.s. (260×)

DESCRIPTION: A nerve is composed of the axonal processes of numerous neurons. Myelin surrounds many of the axons. Connective tissues wrap the axons: the endoneurium surrounds each axon; the perineurium bundles groups of axons into fascicles; the epineurium covers the entire nerve.

LOCATION: Found throughout the body carrying sensory and motor innervation from/to the periphery.

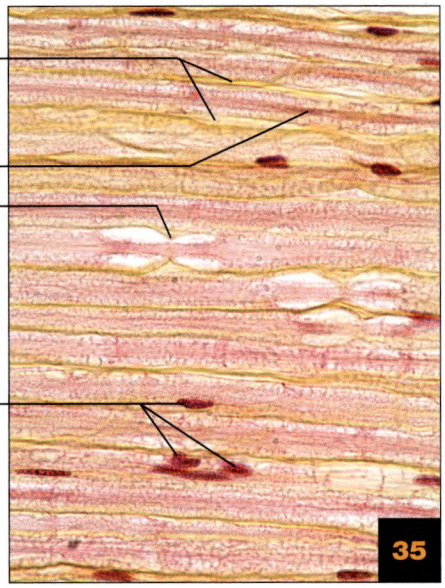

PLATE 35 Peripheral nerve, l.s. (830×)

DESCRIPTION: Longitudinal section through a nerve showing multiple axons surrounded by myelin sheaths; Nodes of Ranvier, pinching of the myelin sheath that indicates the boundary between adjacent Schwann cells; and Schwann cell nuclei.

LOCATION: Found throughout the body innervating body organs.

SELECT ORGANS

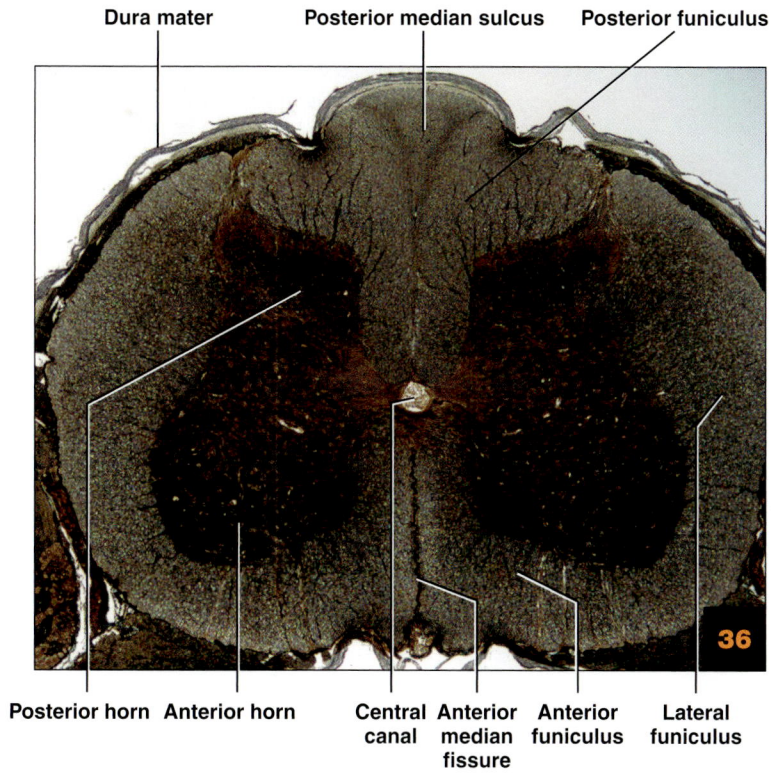

PLATE 36 Spinal cord c.s. through lumbar region (18×)

DESCRIPTION: The hollow central canal is surrounded by a butterfly-shaped region of gray matter forming the anterior and posterior horns (stained brown in this section) that contain neuronal cell bodies; short, non-myelinated interneurons; and neuroglia. The external white matter, the funiculi, (stained gray) contains myelinated axons that make up the ascending and descending spinal tracts.

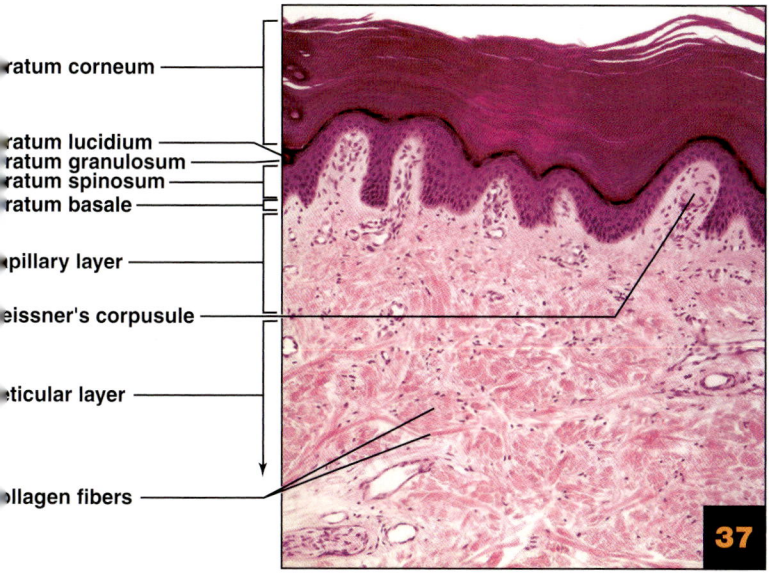

PLATE 37 Thick skin showing epidermal and dermal regions (85×)

DESCRIPTION: The epidermis, the dark pink region, is composed of the stratum basale, stratum spinosum, stratum granulosum, stratum lucidium, and the thick stratum corneum. Deep to the epidermis is the papillary layer of the dermis, composed of loose areolar connective tissue. Note the Meissner's corpuscle, a touch receptor, in the dermal papilla. The deepest layer of the dermis, the reticular layer, is composed of dense irregular connective tissue. Note the greater density of collagen fibers (stained pink) in this layer.

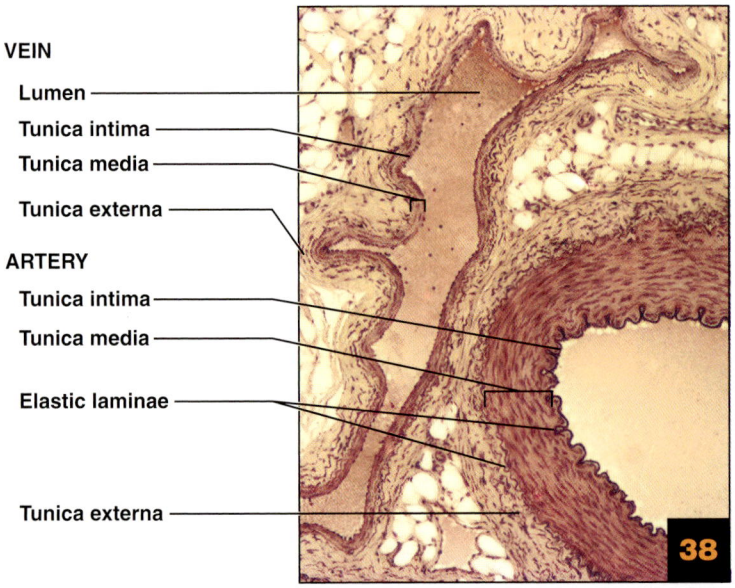

PLATE 38 Muscular artery and vein (80×)

DESCRIPTION: The vessel on the right, the artery, shows a wavy tunica intima resulting from the internal elastic lamina just deep to this layer. The thick tunica media is composed of multiple cell layers. Another layer of wavy elastic tissue, the external elastic lamina, is located outside of the tunica media. In the vein on the left, the lumen is irregular in shape, there are no elastic laminae, and the tunica media is only a few cells in thickness.

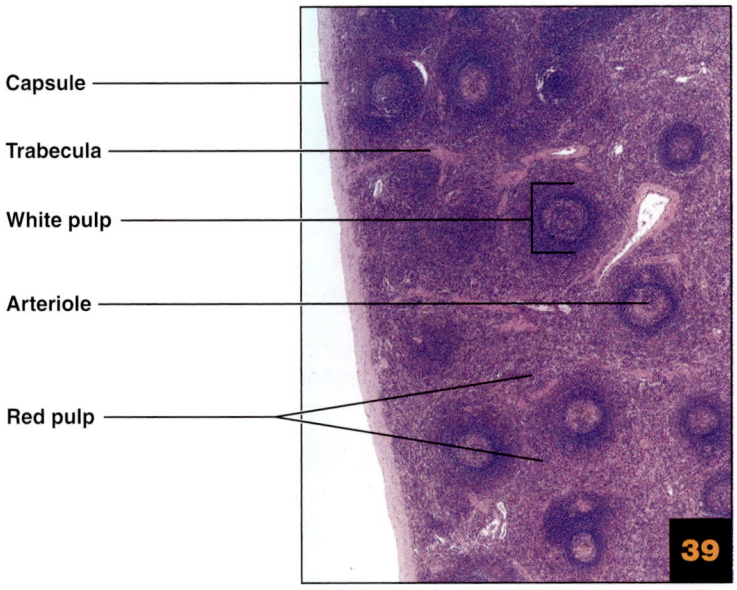

PLATE 39 Spleen (17×)

DESCRIPTION: The outer capsule of the spleen is reticular connective tissue. Trabeculae, extensions of this tissue into the deeper portion of the spleen, provide a framework for the organ. Lymphoid tissue, containing B and T lymphocytes surrounding arterial branches forms the white pulp of the spleen. Surrounding these "islands" is the red pulp, composed of both venous sinuses and splenic cords.

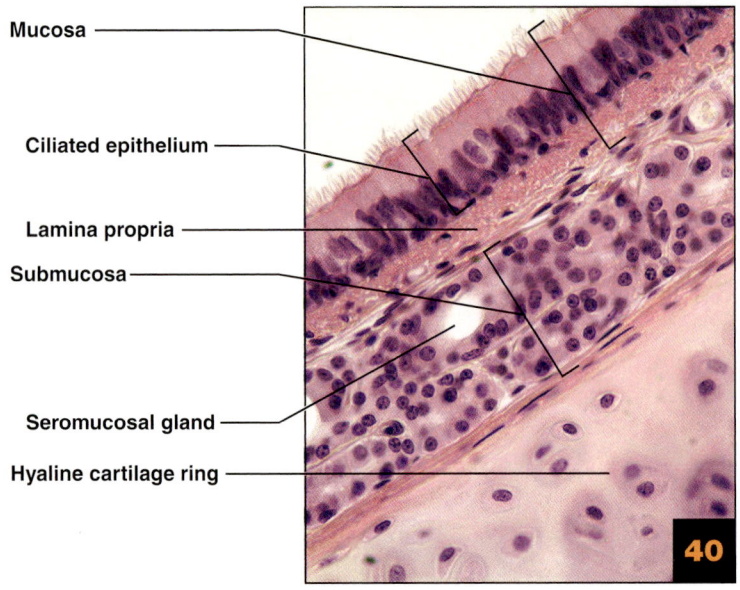

PLATE 40 Trachea (355×)

DESCRIPTION: The mucosa, composed of pseudostratified ciliated columnar epithelium and the underlying lamina propria, lines the trachea. Note the columnar epithelial cells, multiple nuclei, and the distinctive cilia along the apical surface. External to this layer, the submucosa consists of connective tissue and imbedded seromucosal glands that secrete mucous. Cartilagenous rings, composed of hyaline cartilage make up the most external layer seen.

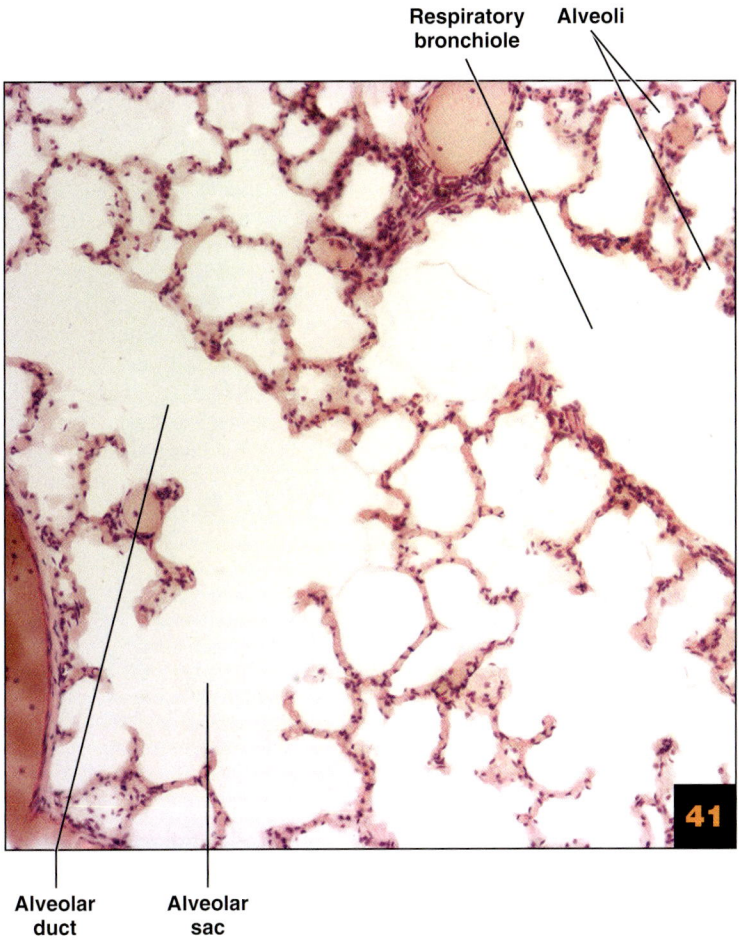

PLATE 41 Lung (120×)

DESCRIPTION: This section through the lung shows the structures of the respiratory zone. The respiratory bronchiole, lined with simple cuboidal epithelium, has alveoli outpocketing from its wall. Alveolar ducts lead to alveolar sacs, terminal clusters of alveoli. Simple squamous epithelium forms the alveolar walls.

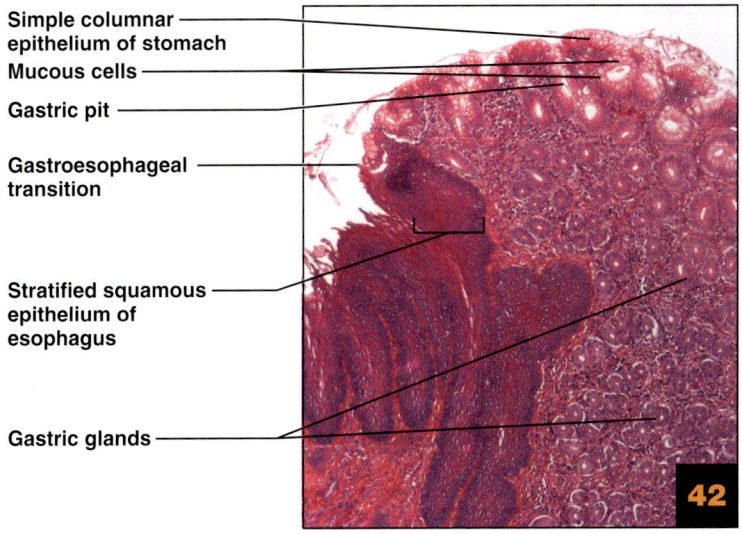

PLATE 42 Gastroesophageal junction (120×)

DESCRIPTION: The epithelial tissue changes abruptly at the gastroesophageal junction, from stratified squamous epithelium in the esophagus (on the left side of the image) to simple columnar epithelium of the stomach (top of the image). The gastric pits and gastric glands of the mucosal layer of the stomach, also composed of simple columnar epithelium, are apparent.

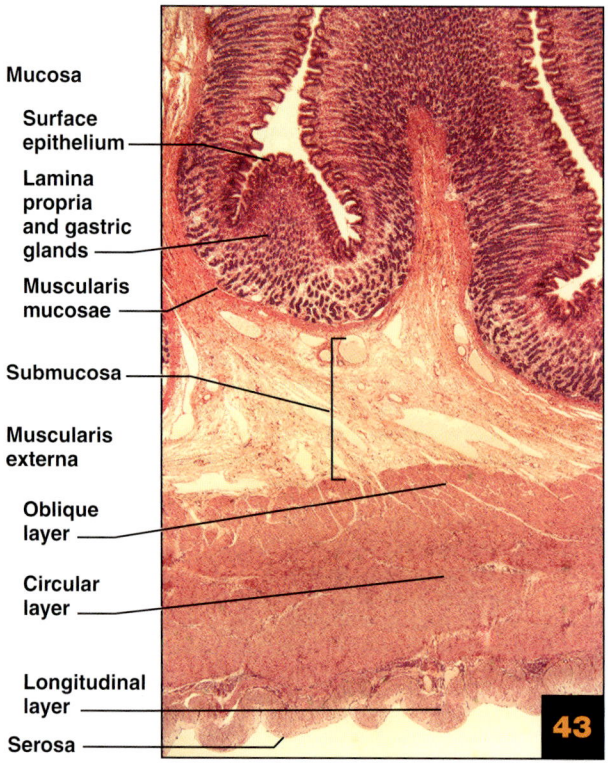

PLATE 43 Stomach, l.s. (17×)

DESCRIPTION: The mucosa is composed of the simple columnar epithelium forming the surface epithelium and the gastric glands, the lamina propria surrounding the glandular epithelium, and the muscularis mucosae, a thin layer of smooth muscle. The submucosa is a moderately dense connective tissue. The muscularis externa is three layers of smooth muscle: a deep oblique layer (difficult to distinguish at this magnification), a middle circular layer, and an external longitudinal layer. The serosa, the outermost layer, is loose areolar connective tissue and mesothelium.

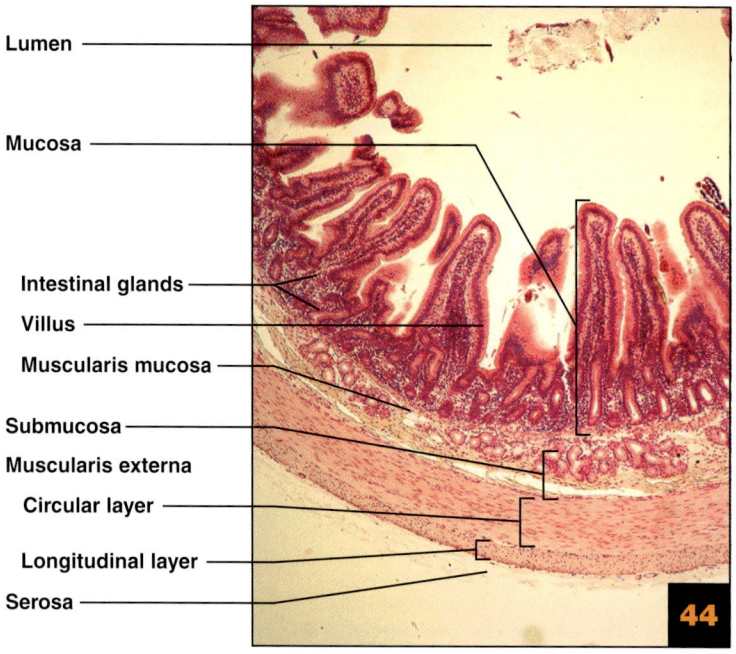

PLATE 44 Small intestine, c.s. through duodenum (37×)

DESCRIPTION: The four layers of the wall of the small intestine are shown: the innermost layer, the mucosa, showing villi extending into the lumen, intestinal glands, and the muscularis mucosae; the submucosa, connective tissue imbedded with duodenal glands (only in the duodenum); the two layers of the muscularis externa, inner circular and outer longitudinal layers; and the outermost serosa composed of loose areolar connective tissue and simple squamous epithelium (mesothelium).

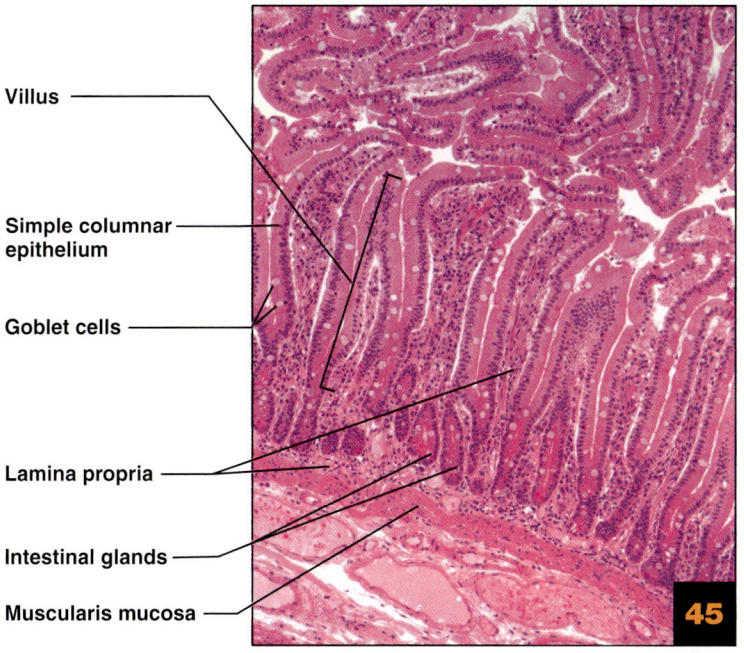

PLATE 45 Mucosal layer of small intestine, from the jejunum (76×)

DESCRIPTION: Details of the mucosal layer are shown. Villi, lined with simple columnar epithelium and goblet cells, extend into the intestinal lumen; intestinal glands at the base of the villi produce digestive secretions; muscularis mucosae, a smooth muscle layer, forms the outermost layer of the mucosa. Refer to Plate 4 for high power view of mucosa. View the microvilli that make up the brush border on the apical surface of the columnar epithelial cells.

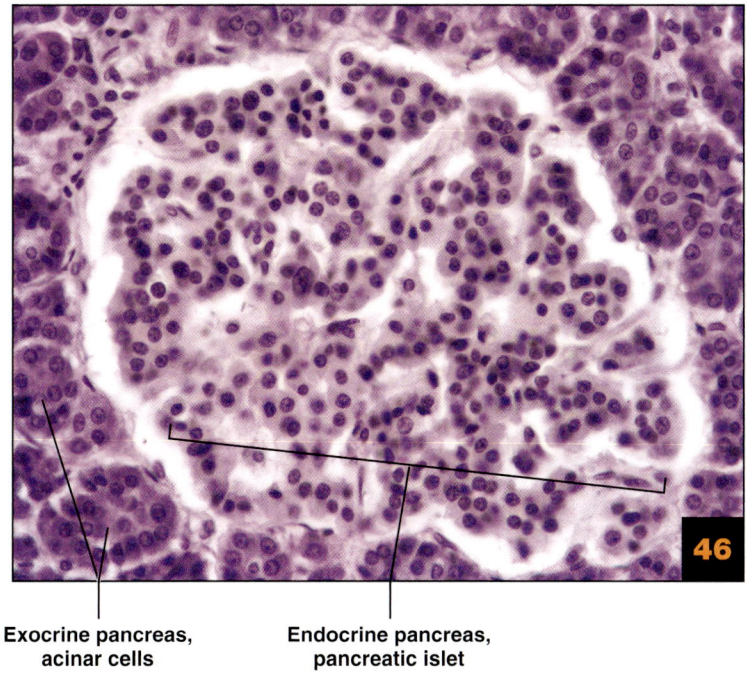

PLATE 46 Pancreas (350×)

DESCRIPTION: The glandular cells in the center of the field make up the pancreatic islet, the endocrine portion of the pancreas that produces insulin (beta cells), glucagon (alpha cells), and somatostatin (delta cells). These hormones are secreted into capillaries surrounding these cells. The exocrine pancreas, the acinar cells surround the islet. These cell produce digestive secretions that empty into the duodenum via ducts.

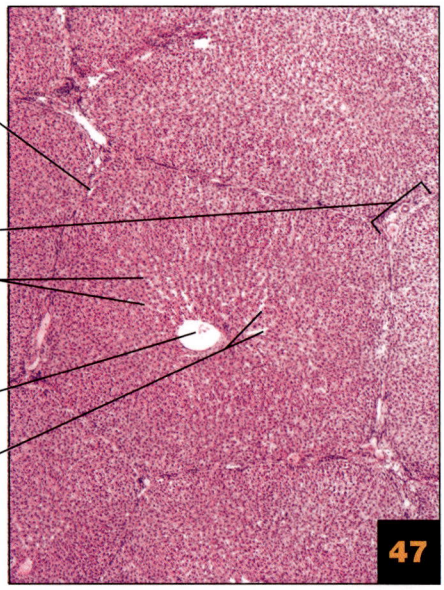

PLATE 47 Pig liver (34×)

DESCRIPTION: Liver lobules are not distinct in humans. Pig liver is shown here to illustrate the lobular structure. A central vein runs through the middle of the lobule. Radiating outward are cords containing hepatocytes. Blood sinusoids are located between the hepatic cords. Hepatic triads, composed of branches of the hepatic artery, hepatic portal vein, and the bile duct, are found at the corners of the lobule. Individual lobules are separated by connective tissue septa.

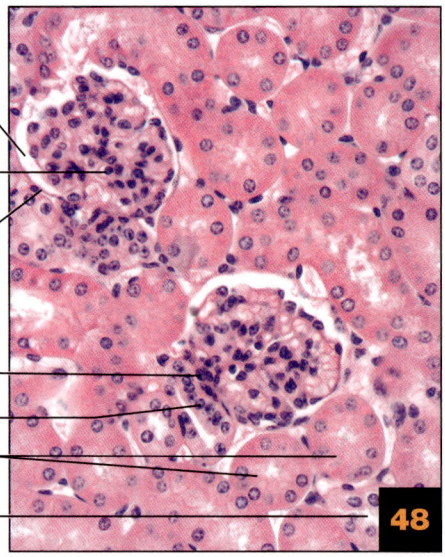

PLATE 48 Renal cortex (270×)

DESCRIPTION: Two renal corpuscles are shown surrounded by renal tubules in cross section. Simple squamous epithelium makes up the outer portion of the glomerular capsule and simple cuboidal epithelium forms the portions of the renal tubules shown. The proximal convoluted tubules appear thicker due to the microvilli extending from their apical surface. The corpuscle on the right shows the elongated macula densa cells and juxtaglomerular cells of the juxtaglomerular apparatus.

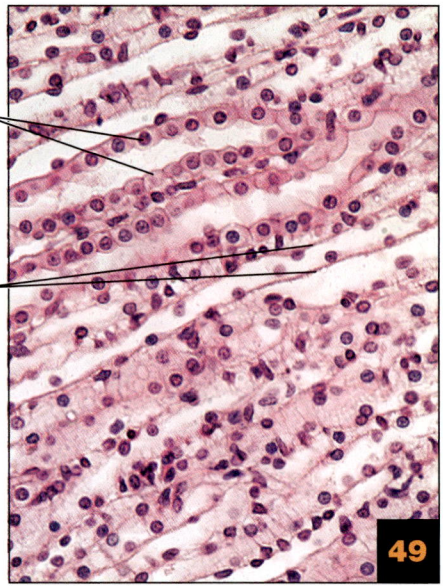

PLATE 49 Renal medulla (350×)

DESCRIPTION: Longitudinal section through renal tubules in the renal medulla. Simple cuboidal epithelium lines the collecting ducts and the thick segments of the loop of Henle (part of the ascending limb). Simple squamous epithelium forms the thin segments of the loop of Henle.

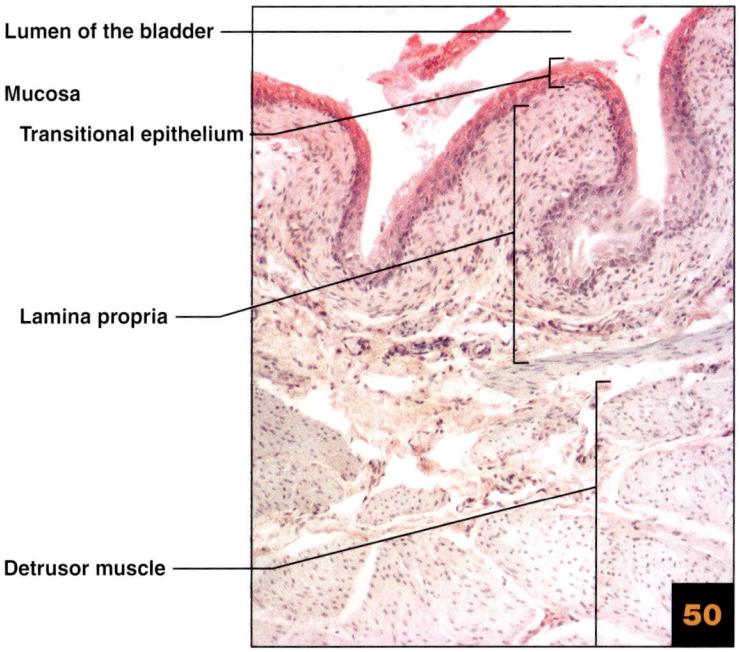

PLATE 50 Urinary bladder (97×)

DESCRIPTION: The transitional epithelium lining the urinary bladder and the lamina propria, composed of loose areolar connective tissue, together make up the mucosa of the bladder. The thick smooth muscle layer, the detrusor muscle, is made up of three indistinct layers. The outer layer, the adventitia, is not shown here.

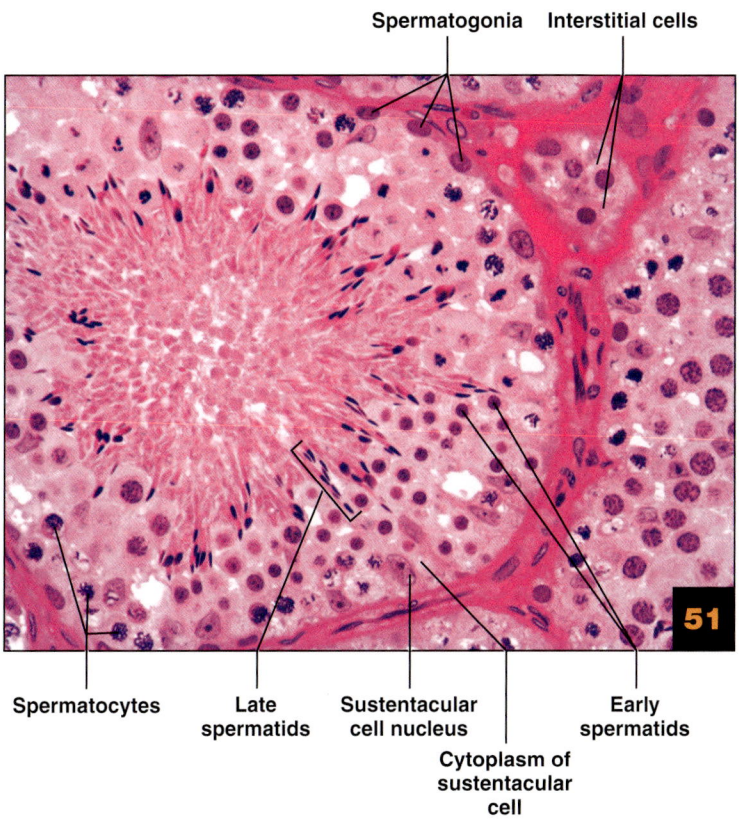

PLATE 51 Testes (340×)

DESCRIPTION: This cross section through a seminferous tubule shows spermatogenic cells embedded within columnar sustentacular cells. Stem cells, spermatogonia, are located at the periphery of the tubule. As cells move through the tubule toward the lumen, sperm are formed. Clusters of cells located between the tubules, interstitial cells, secrete testosterone.

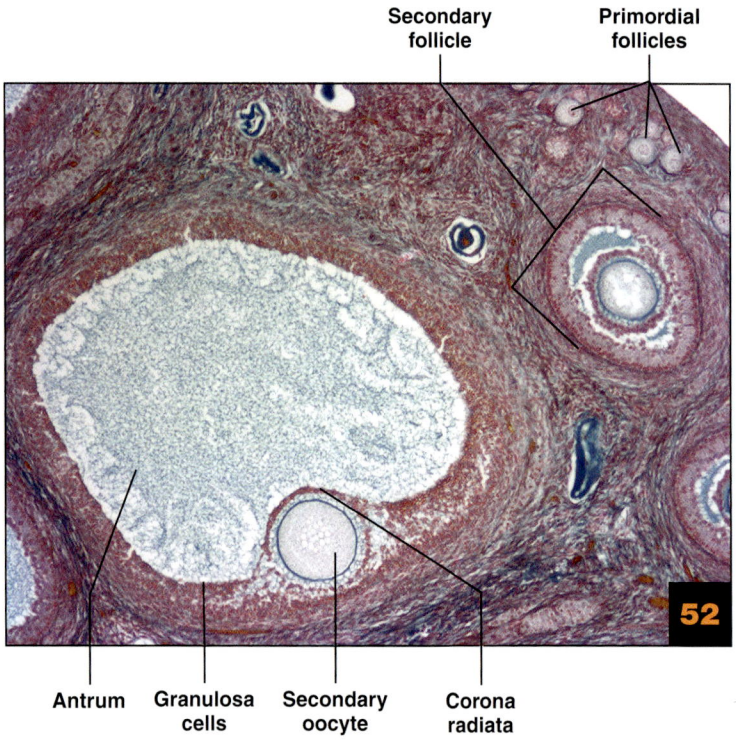

PLATE 52 Ovary (82×, trichrome stain)

DESCRIPTION: This section through an ovary shows follicles in various stages of development. The large follicle is a mature vesicular (Graafian) follicle. It contains a large fluid filled space, the antrum, lined by granulosa cells. Surrounding the oocyte is a ring of granulosa cells called the corona radiata. On the right side of the image, a secondary follicle is shown with the beginning of the antrum forming. The upper right corner contains multiple primordial follicles.

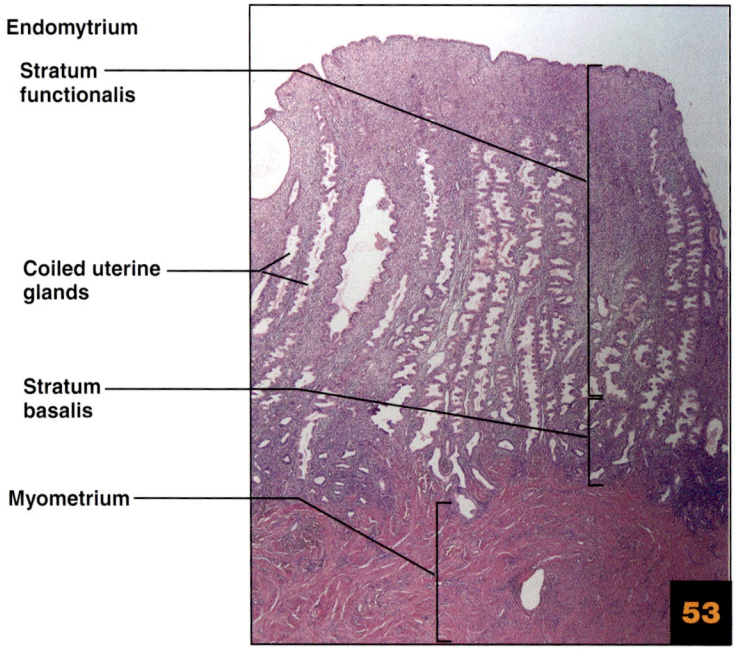

PLATE 53 Uterus, secretory stage (15×)

DESCRIPTION: The lumen of the uterus is at the top of the image. The thick stratum functionalis of the endometrium shows enlarged, coiled uterine glands. The thin stratum basalis forms the base of the endometrium. The myometrium, the smooth muscle layer of the uterine wall, is at the bottom of the image.

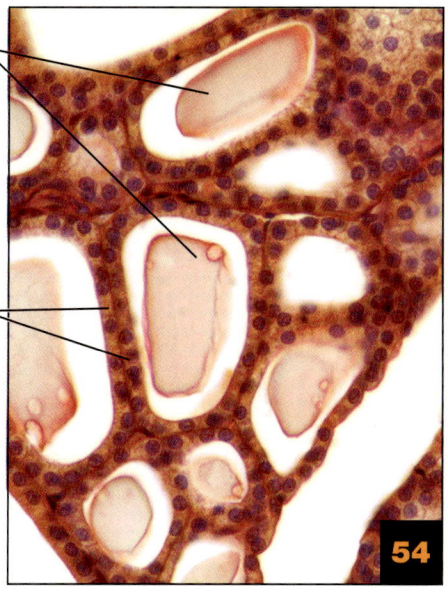

PLATE 54 Thyroid gland (350×)

DESCRIPTION: The thyroid gland is composed of spherical follicles formed by simple cuboidal epithelial cells, follicle cells. The center of each follicle is filled with a gel-like substance called colloid that contains proteins needed for the formation of thyroid hormone. Thyroid hormone is produced by the follicle cells and secreted into the capillaries that surround the follicles. In this image the colloid has pulled away from the follicle cells, an artifact of slide preparation.

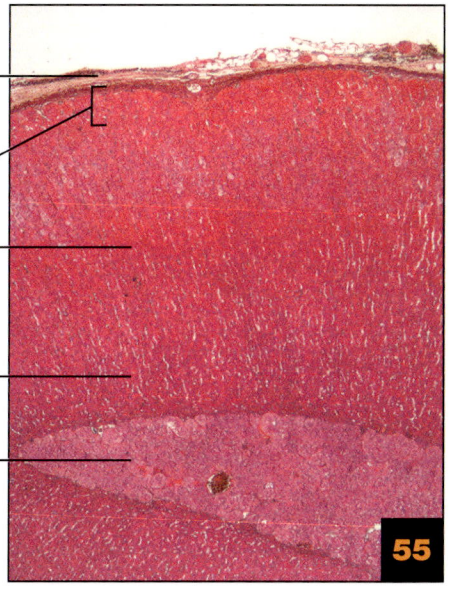

PLATE 55 Adrenal gland, section (35×)

DESCRIPTION: This section through the adrenal gland shows the medulla, light pink oval in the bottom of the image, whose cells secrete epinephrine and norepinephrine; the portions of the cortex: the cells of the thin zona reticularis and the thicker zona fasciculata secrete glucocortidoid hormones, and the cells of the superficial zona glomerulosa secrete mineralocorticoid hormones. The connective tissue forming the adrenal capsule is at the top of the image.

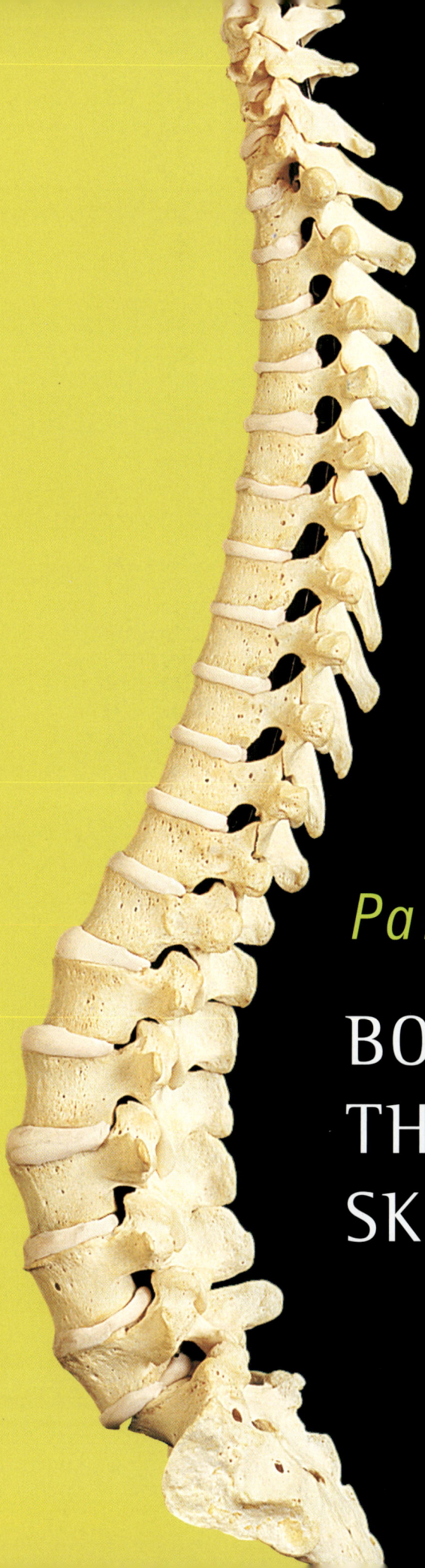

Part II

BONES OF THE HUMAN SKELETON

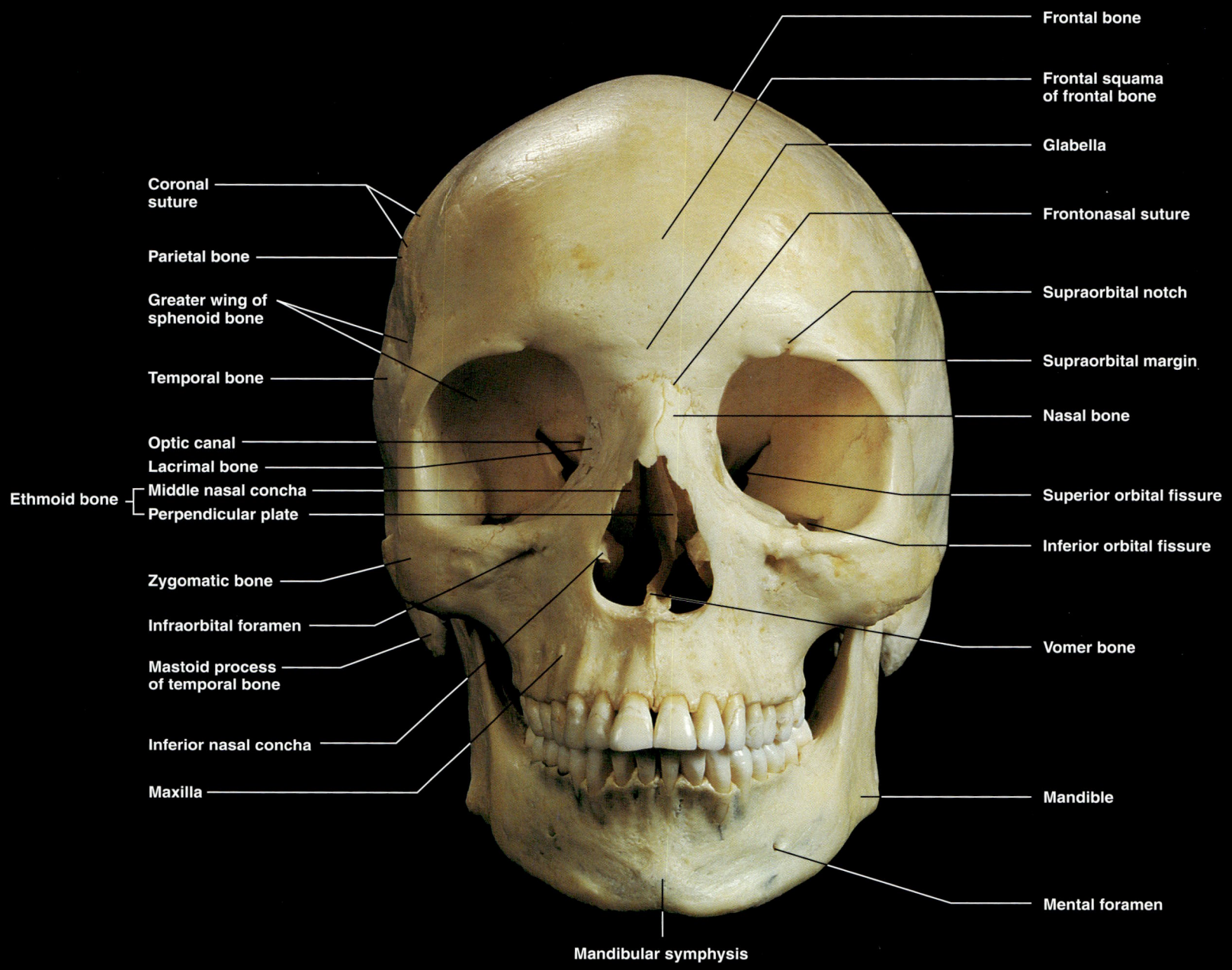

Figure 1 Skull, anterior view.

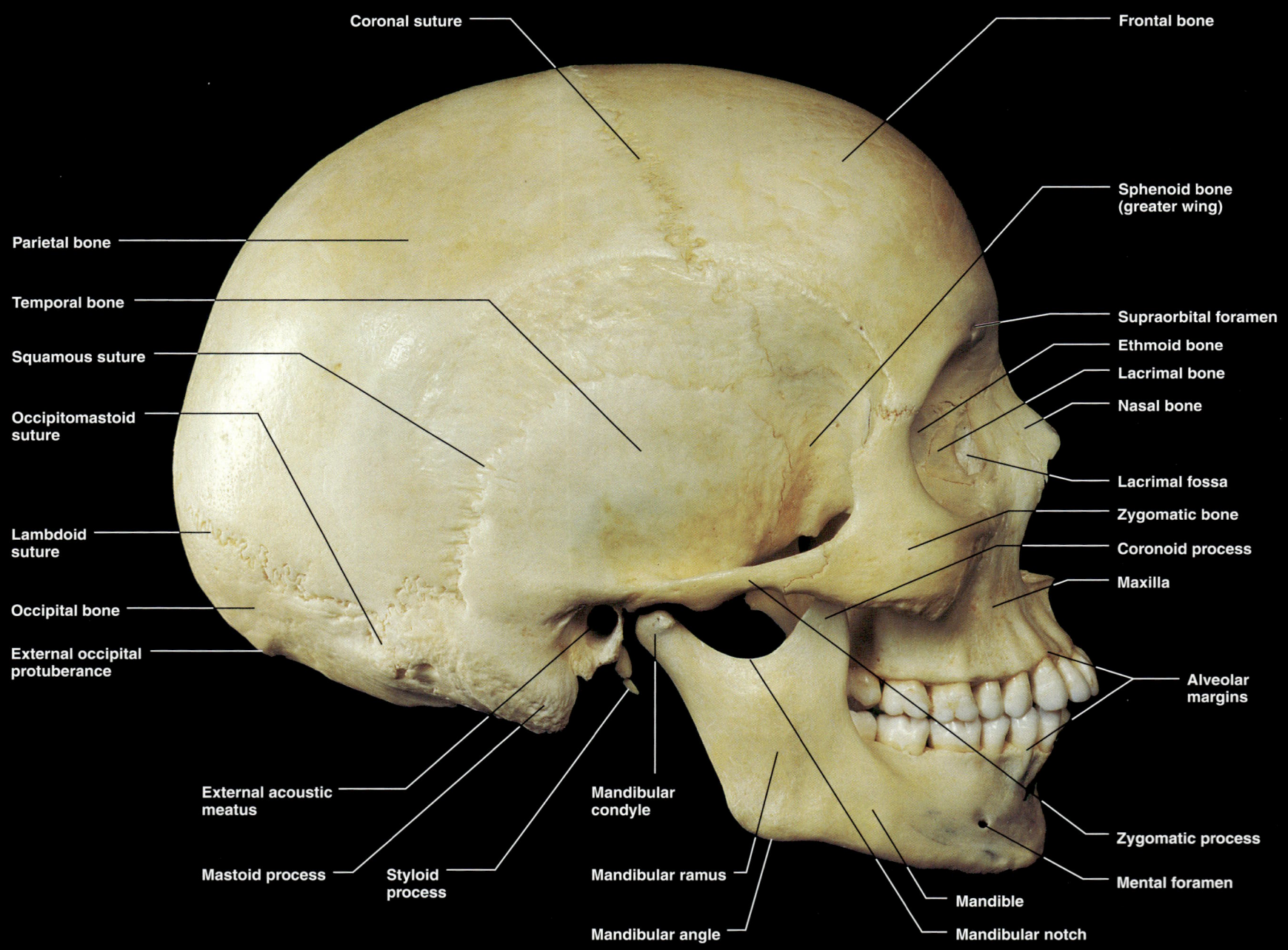

Figure 2 Skull, right external view of lateral surface.

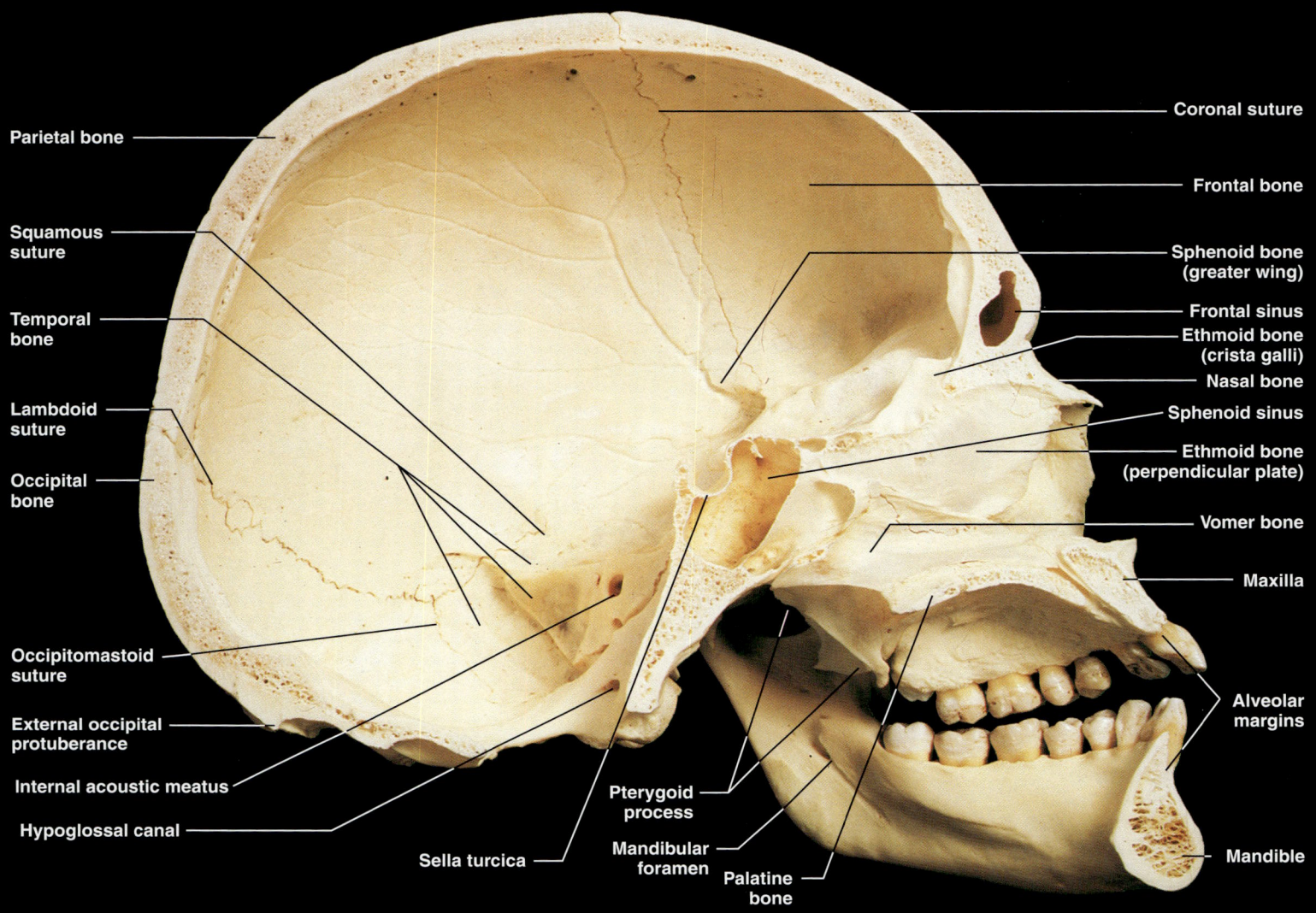

Figure 3 Skull, internal view of left lateral aspect.

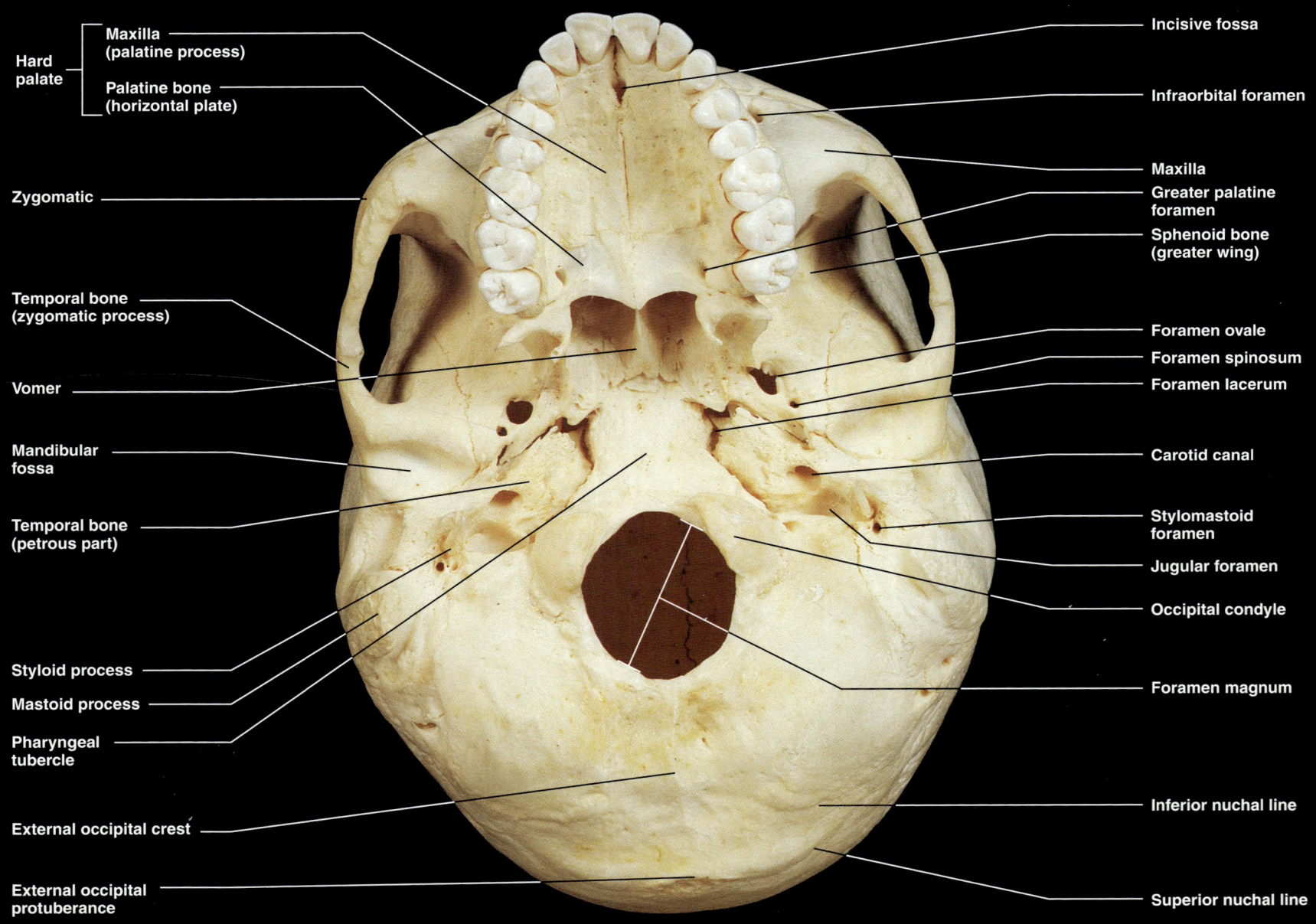

Figure 4 Skull, external view of base.

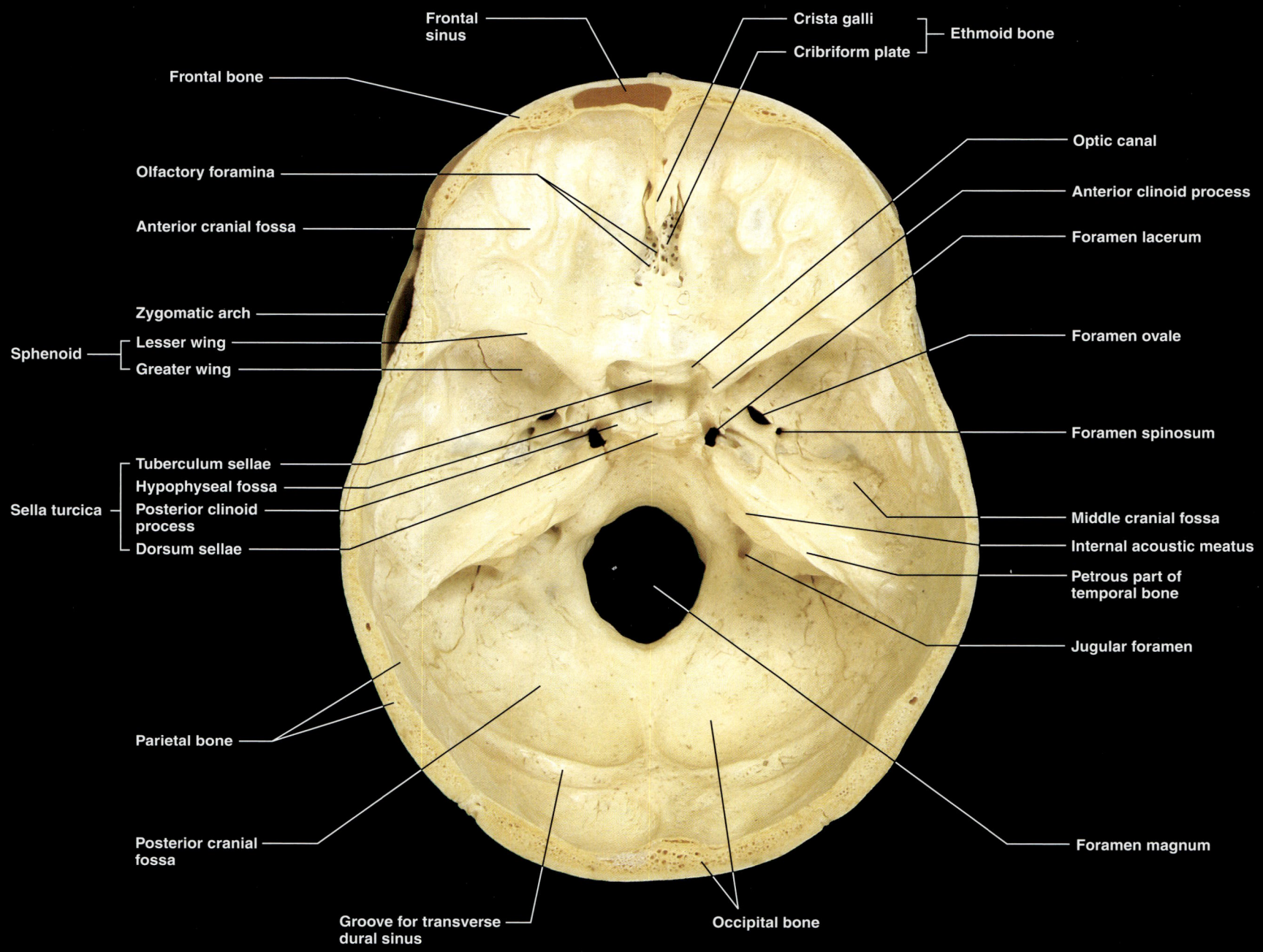

Figure 5 Skull, internal view of base.

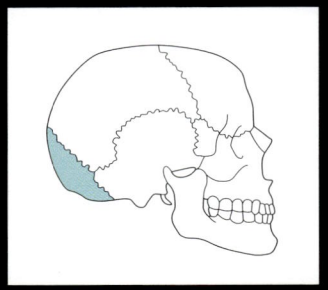

Pharyngeal tubercle

Basioccipital

Occipital condyle

Foramen magnum

Condylar fossa

External occipital crest

Inferior nuchal line

External occipital protuberance

Superior nuchal line

Figure 6 Occipital bone, inferior external view.

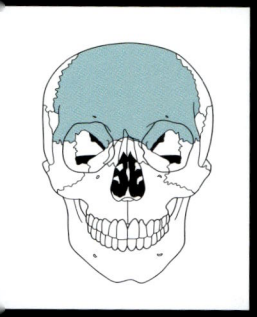

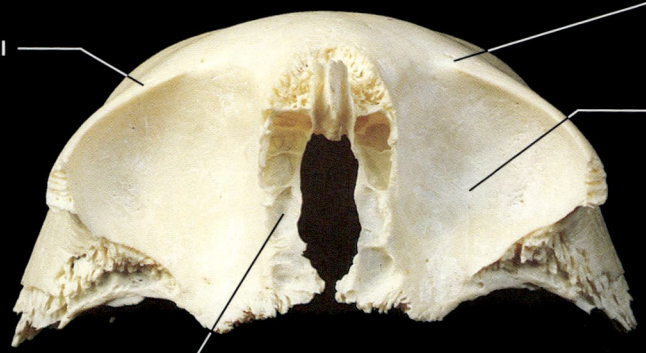

(b) inferior surface

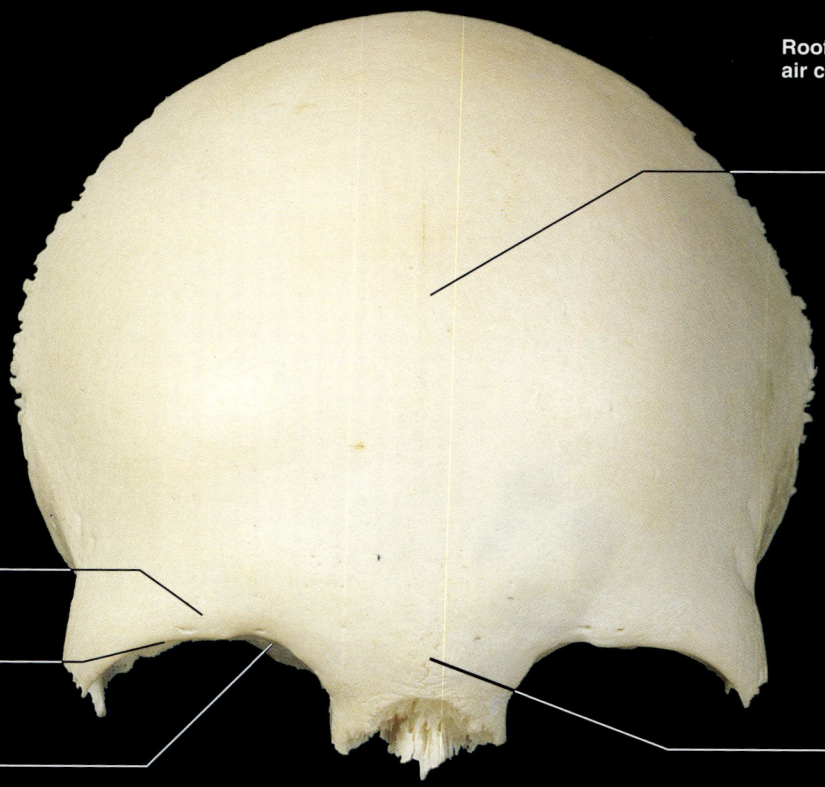

(a) anterior view

Figure 7 Frontal bone.

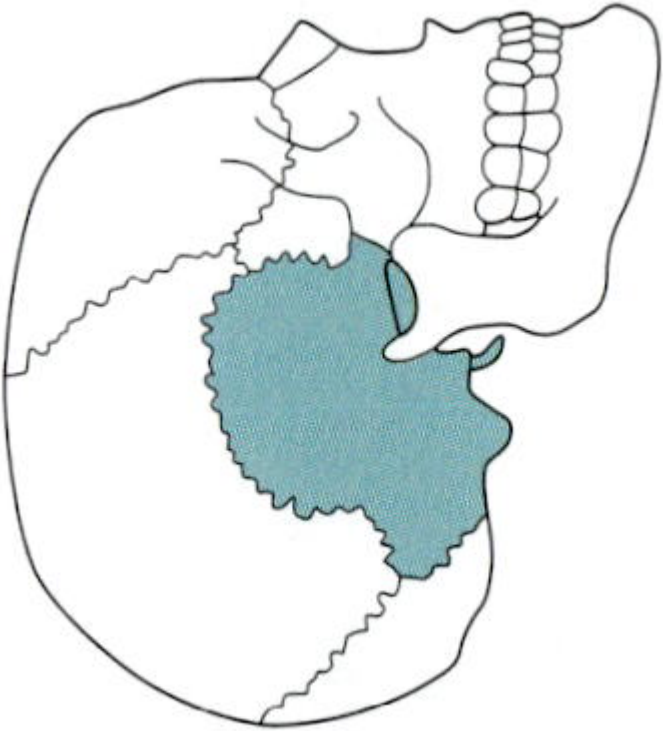

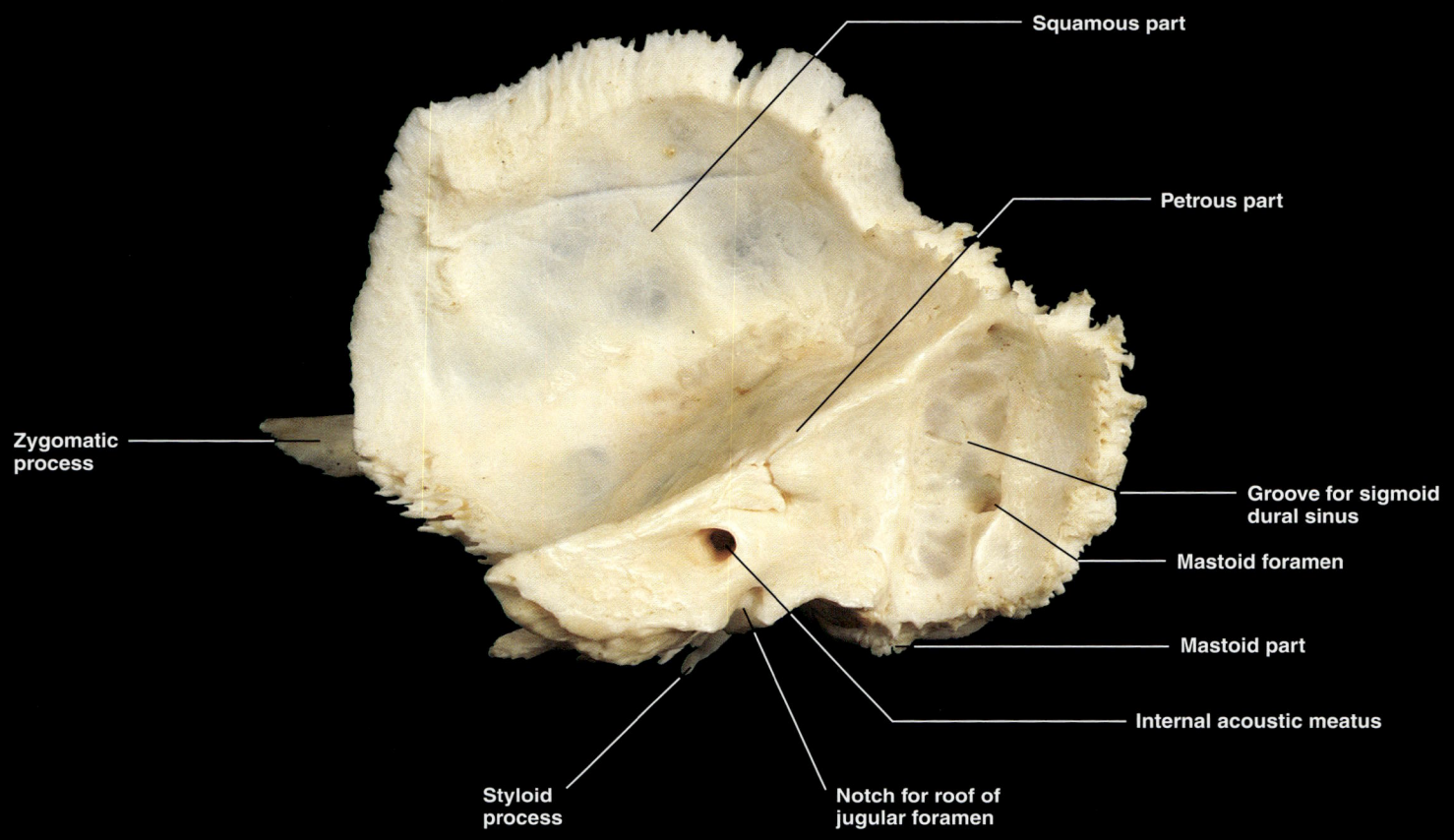

(b) right medial view

Figure 8 Temporal bone.

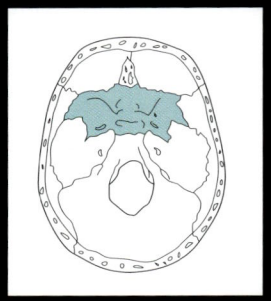

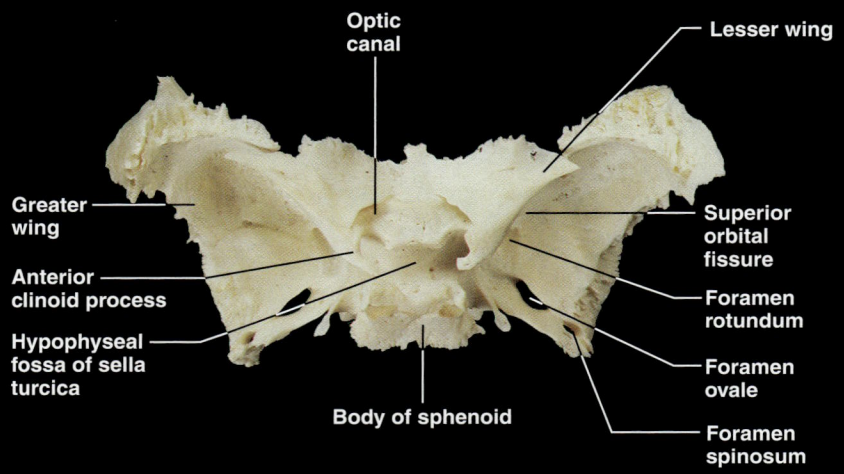

(a) superior view

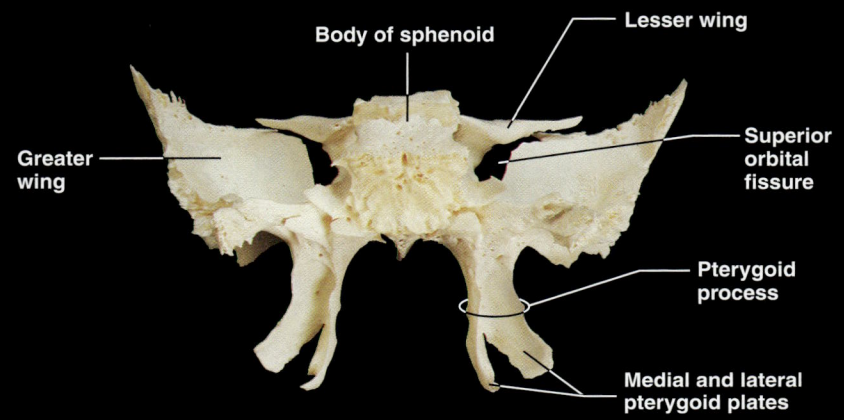

(b) posterior view

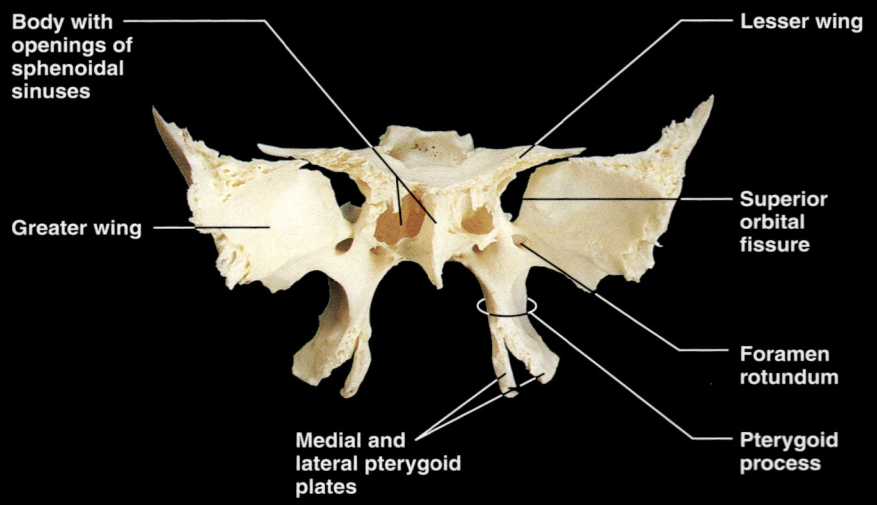

(c) anterior view

Figure 9 Sphenoid bone.

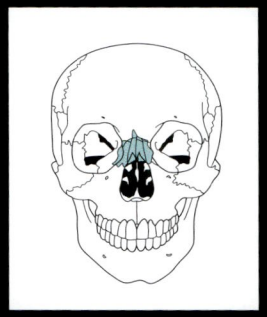

(a) left lateral surface

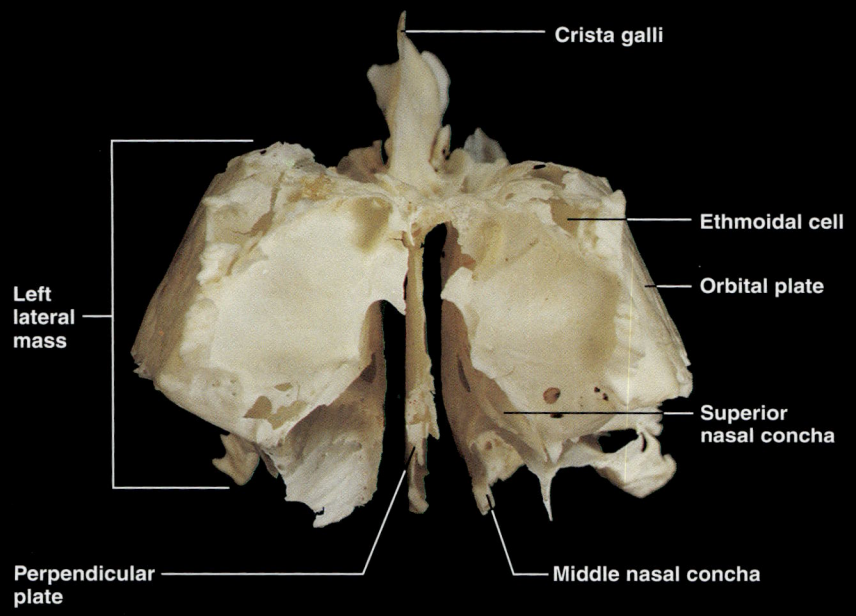

(b) posterior view

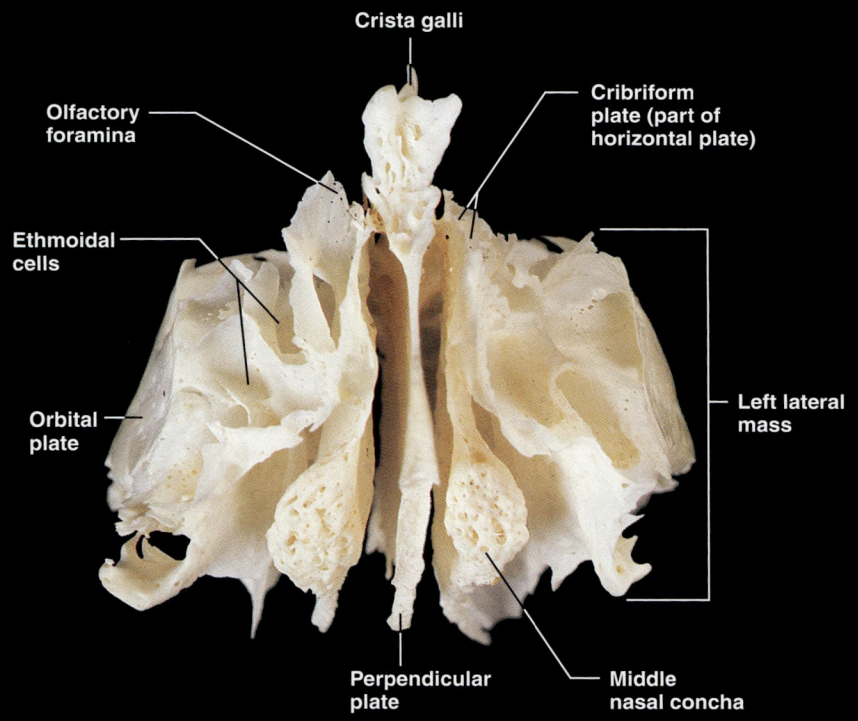

(c) anterior view

Figure 10 Ethmoid bone.

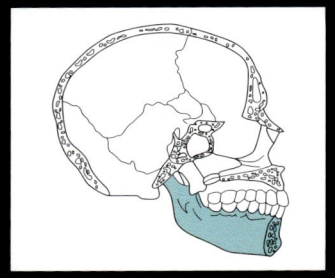

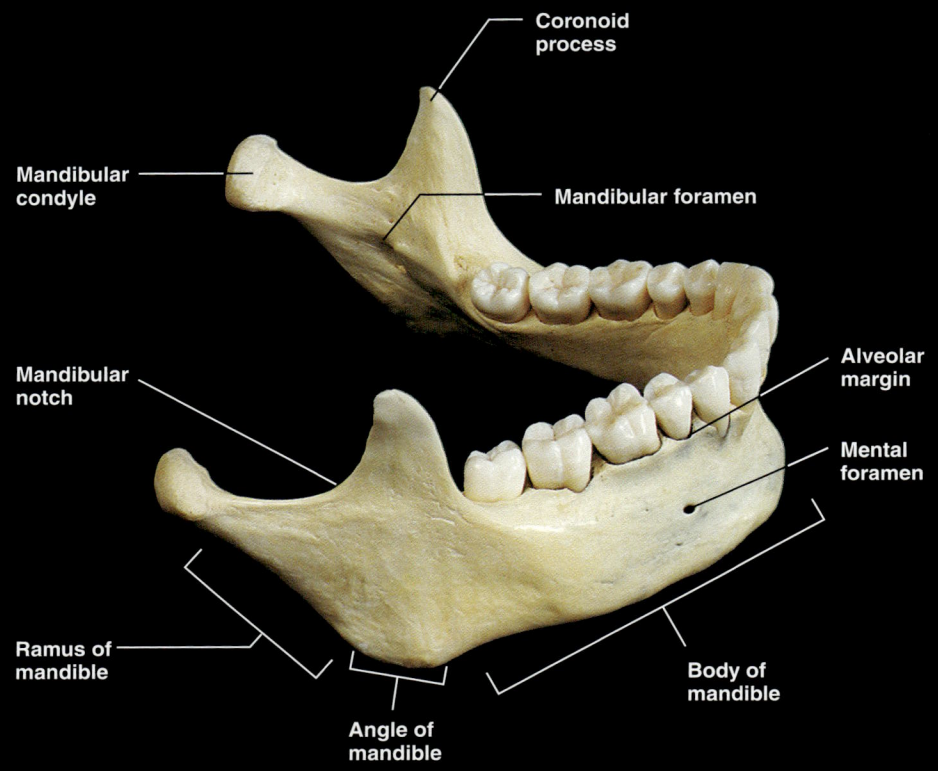

(a) right lateral view

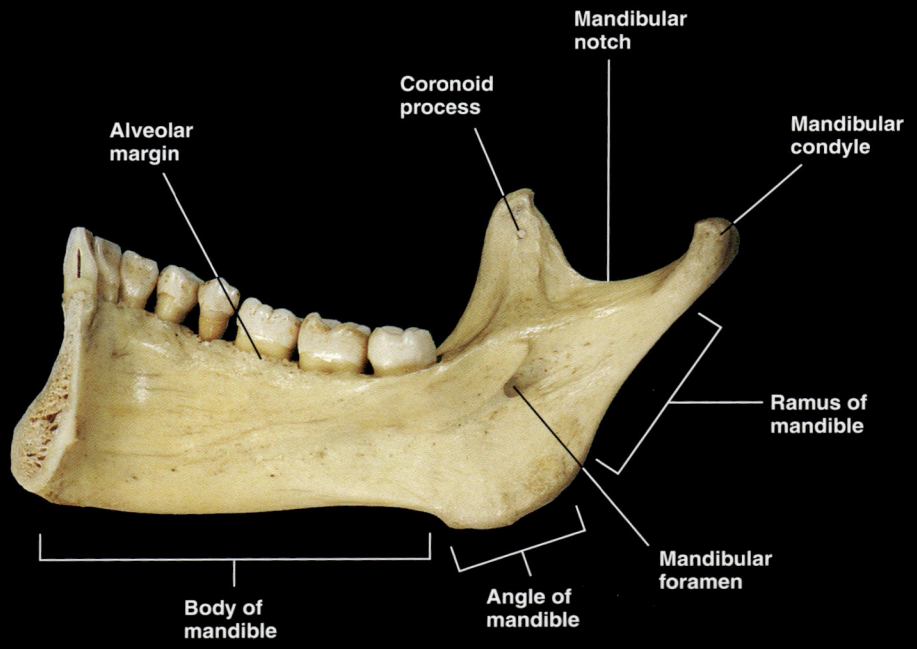

(b) right medial view

Figure 11 Mandible.

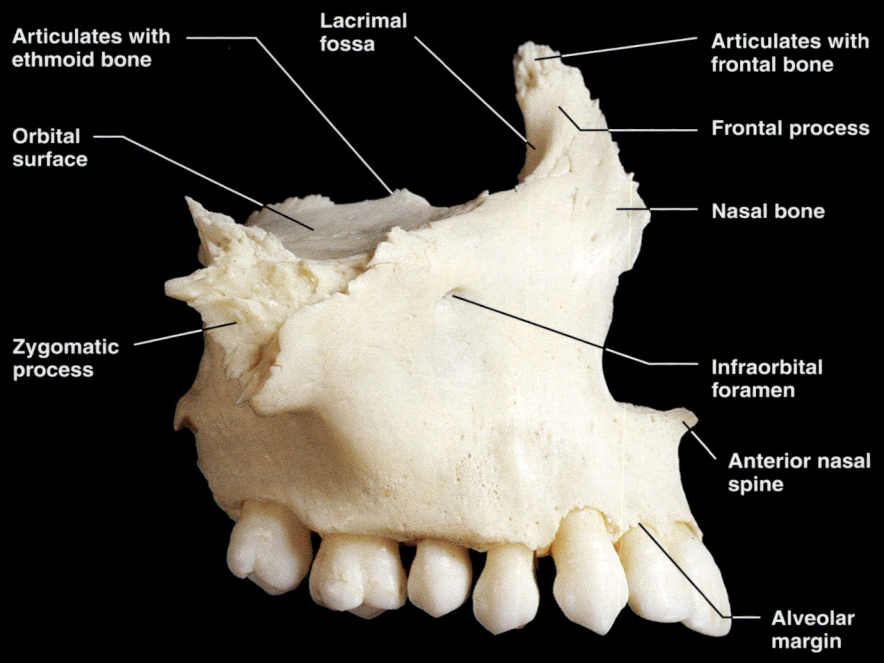

(a) right lateral view

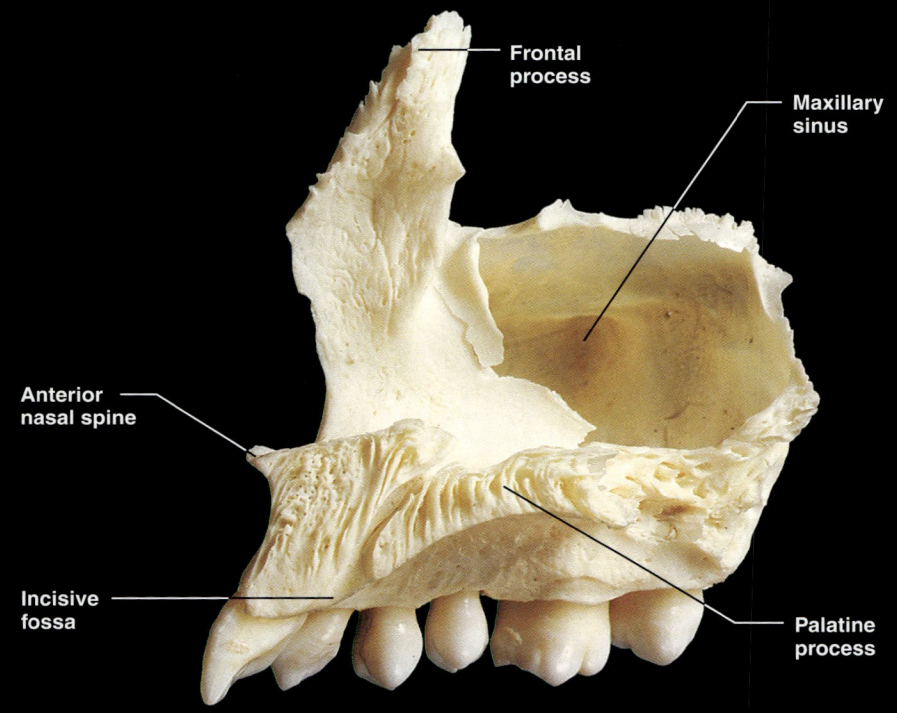

(b) right medial view

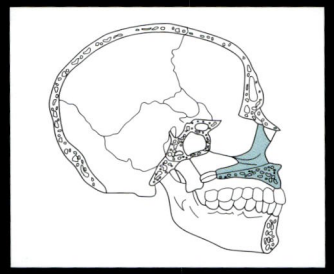

Figure 12 Maxilla.

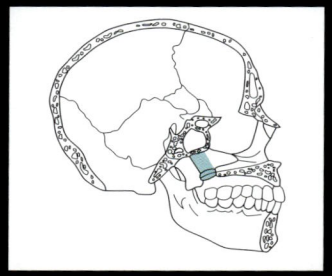

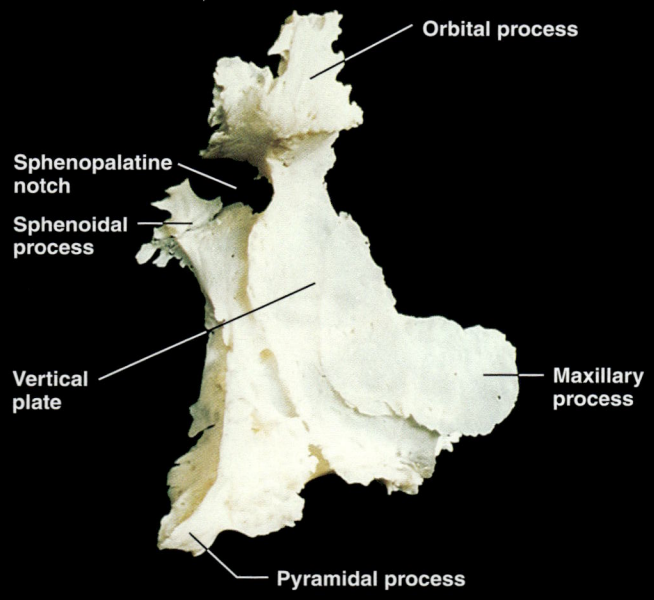

(a) right lateral view

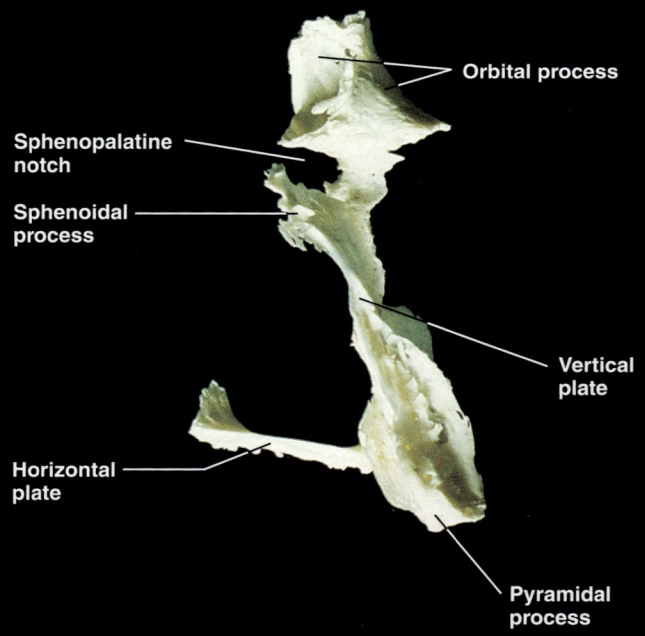

(b) right posterior view

Figure 13 Palatine bone.

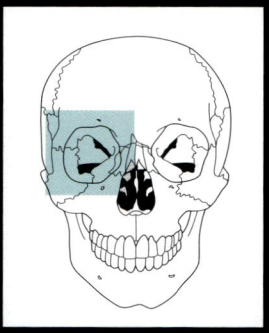

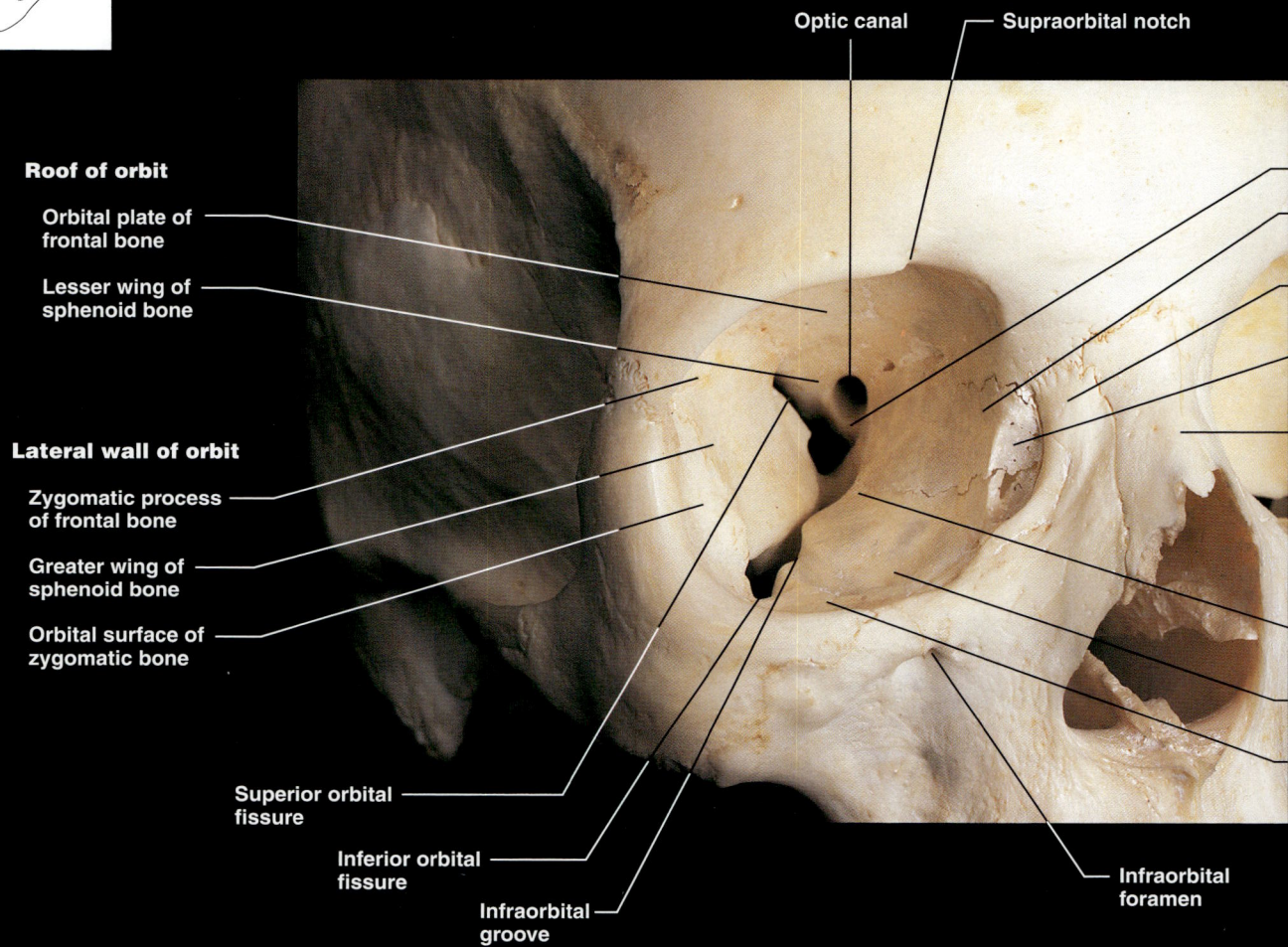
Optic canal
Supraorbital notch
Roof of orbit
Orbital plate of frontal bone
Lesser wing of sphenoid bone
Lateral wall of orbit
Zygomatic process of frontal bone
Greater wing of sphenoid bone
Orbital surface of zygomatic bone
Superior orbital fissure
Inferior orbital fissure
Infraorbital groove
Infraorbital foramen

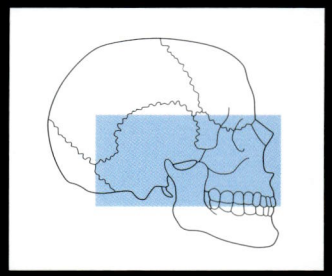

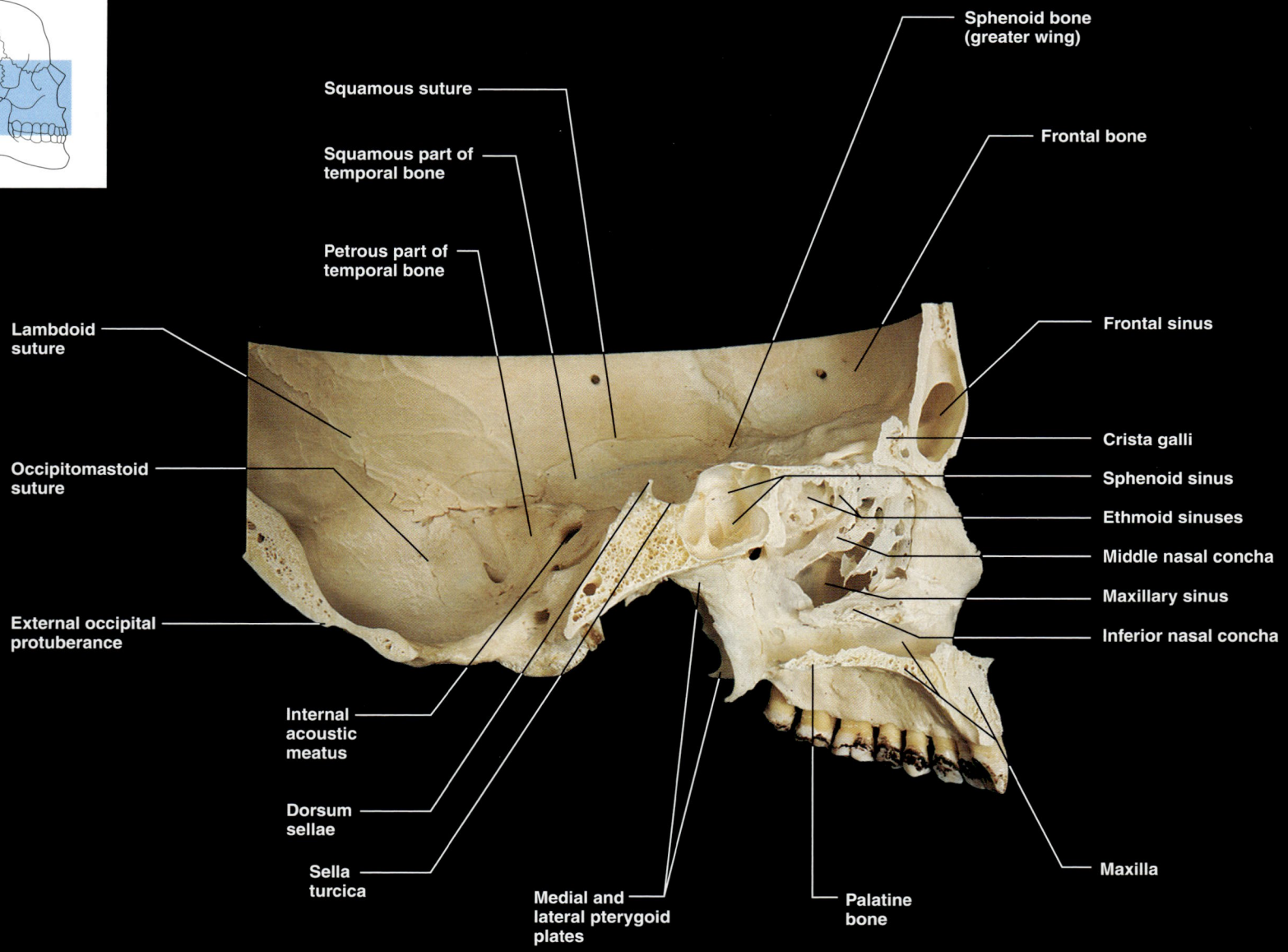

Figure 15 Nasal cavity, left lateral wall.

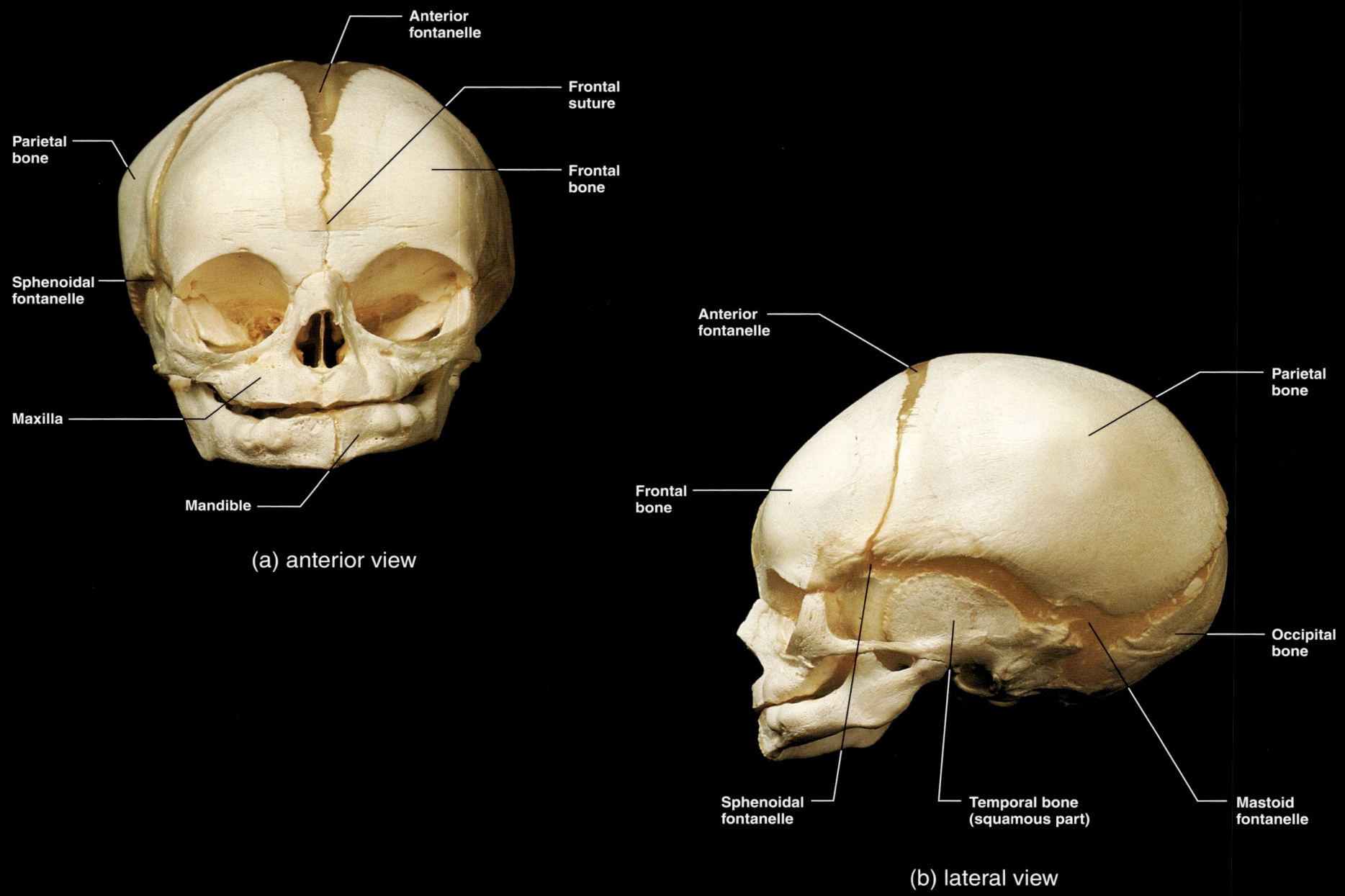

Figure 16 Fetal skull.

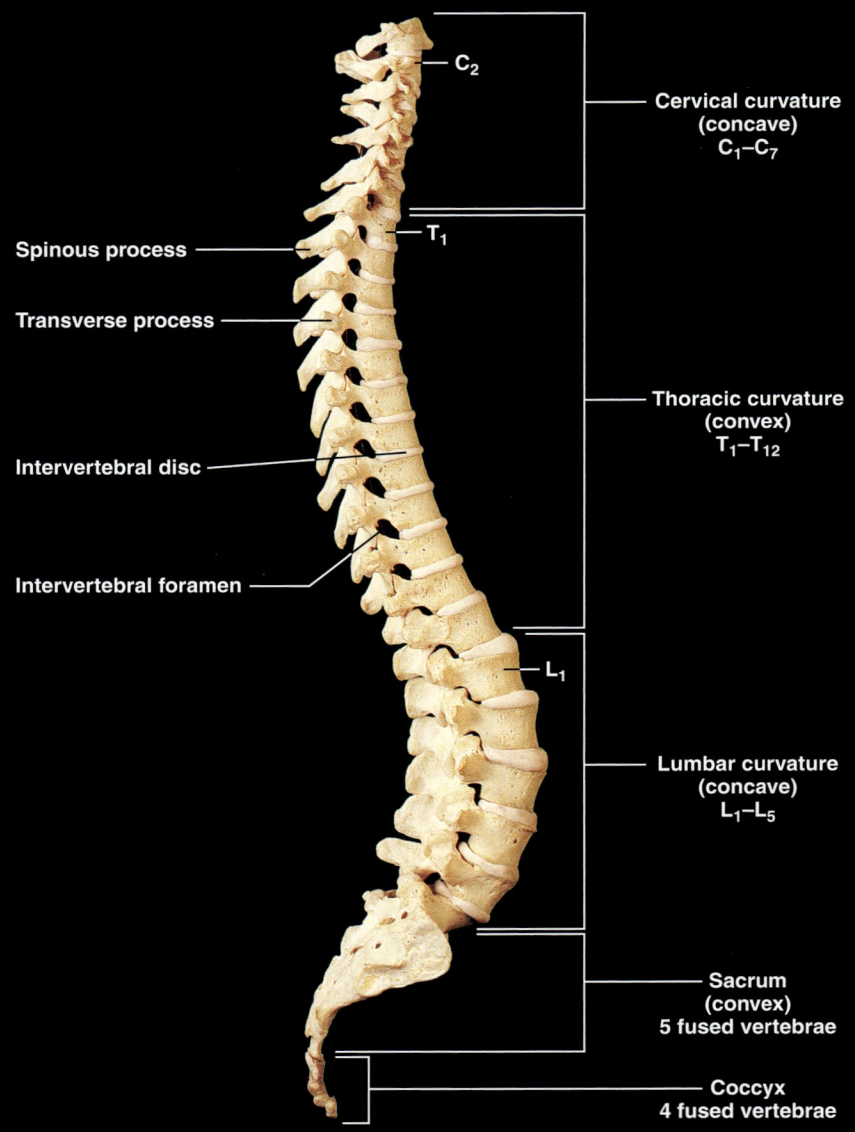

(a) right lateral view

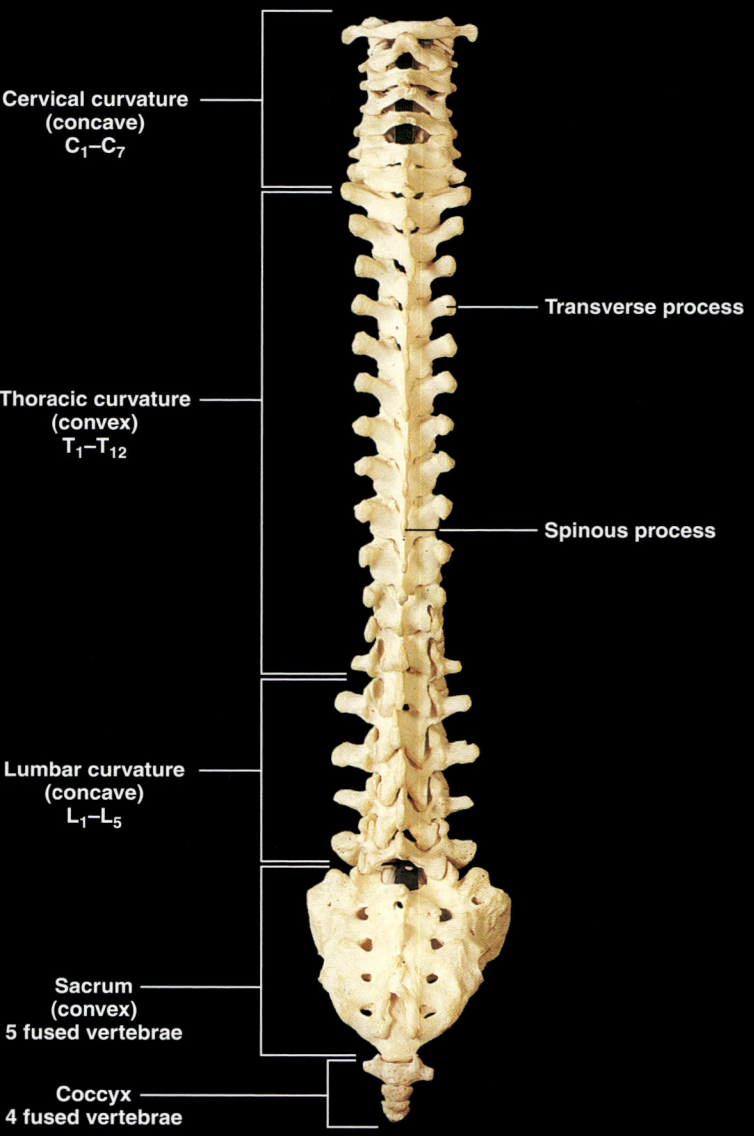

(b) posterior view

Figure 17 Articulated vertebral column.

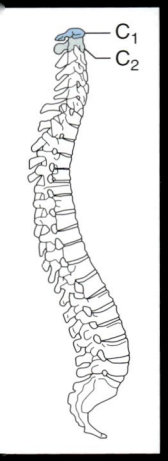

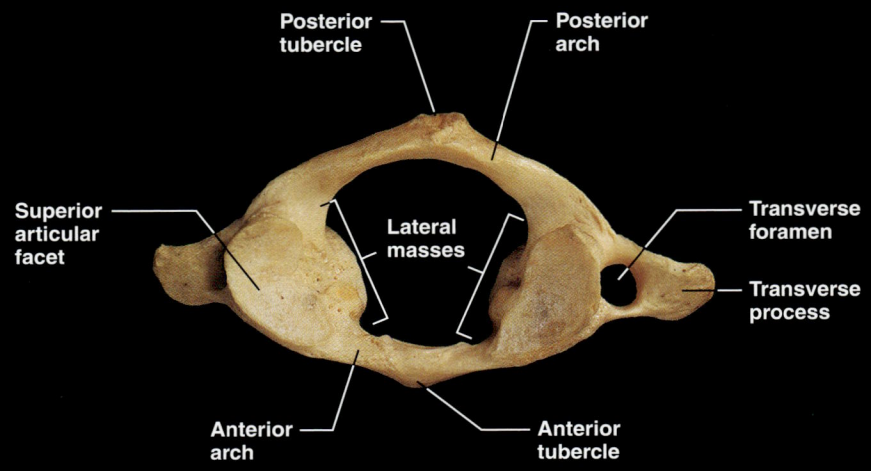

(a) atlas, superior view

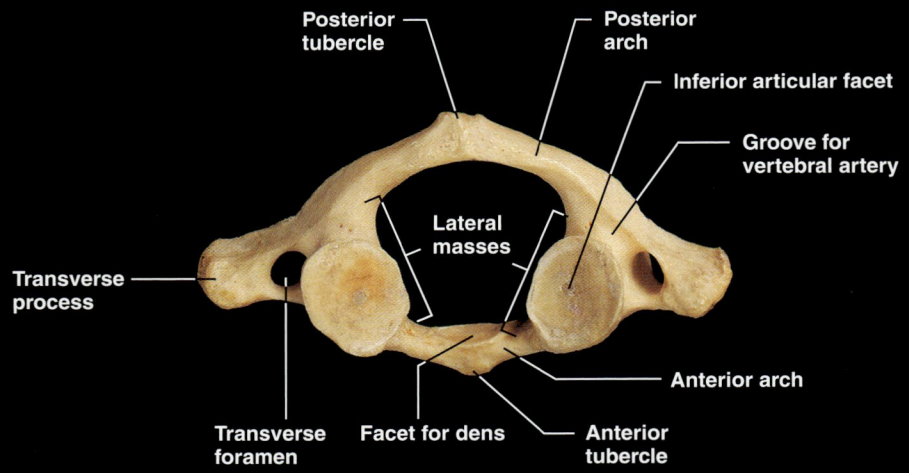

(b) atlas, inferior view

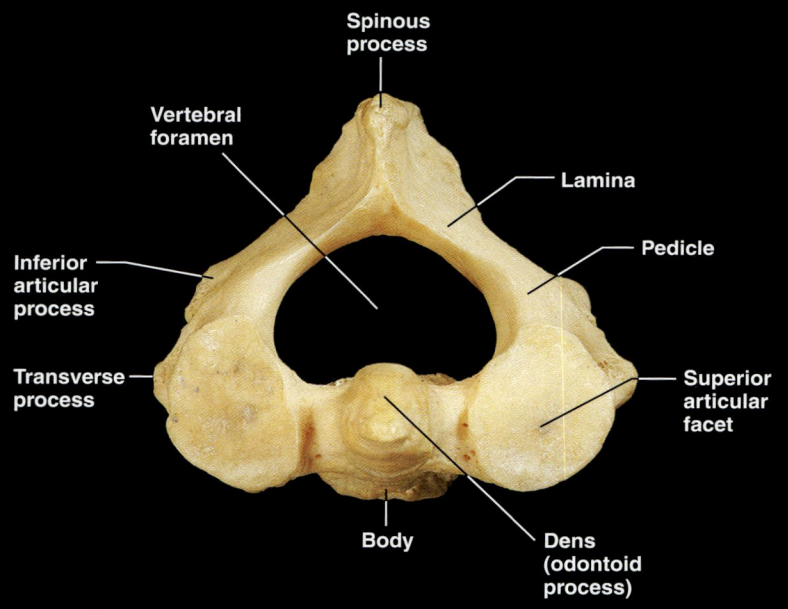

(c) axis, superior view

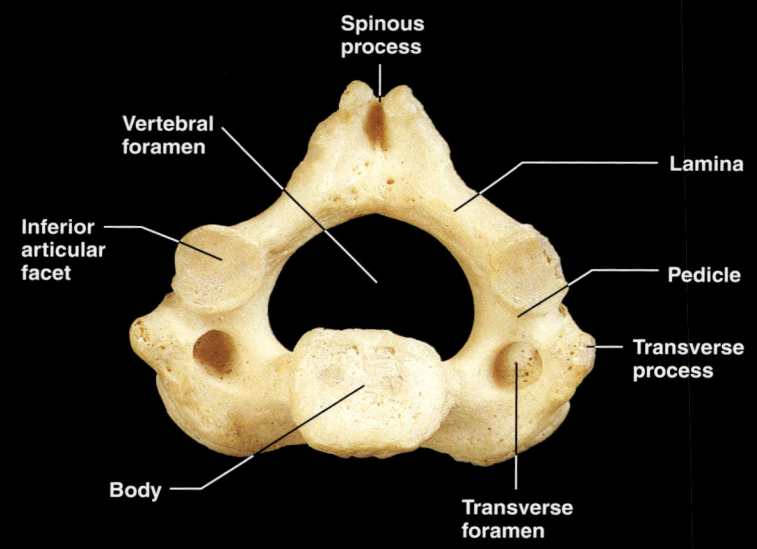

(d) axis, inferior view

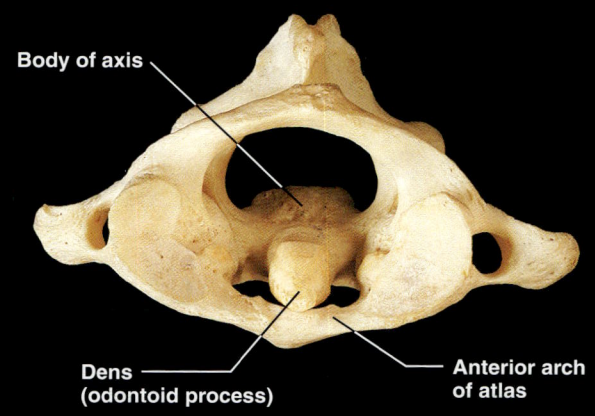

(e) articulated atlas and axis, superior view

Figure 18 Various views of vertebrae C_1 and C_2.

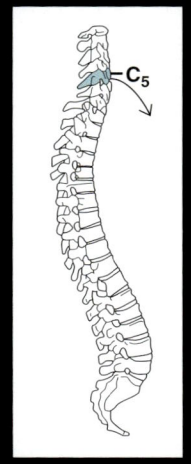

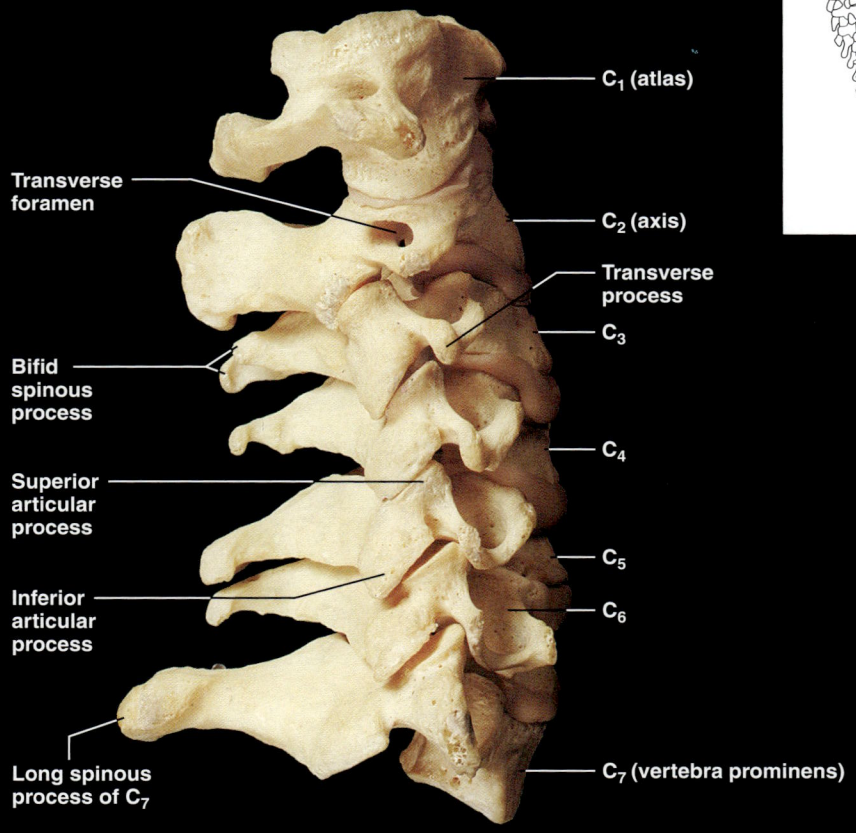

(a) right lateral view of articulated cervical vertebrae

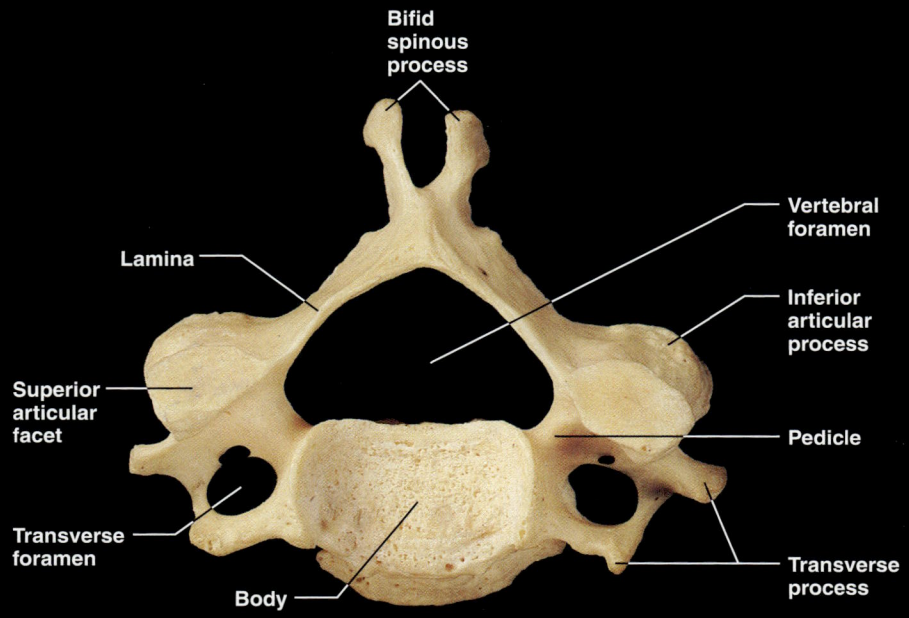

(b) fifth (typical) cervical vertebra, superior view

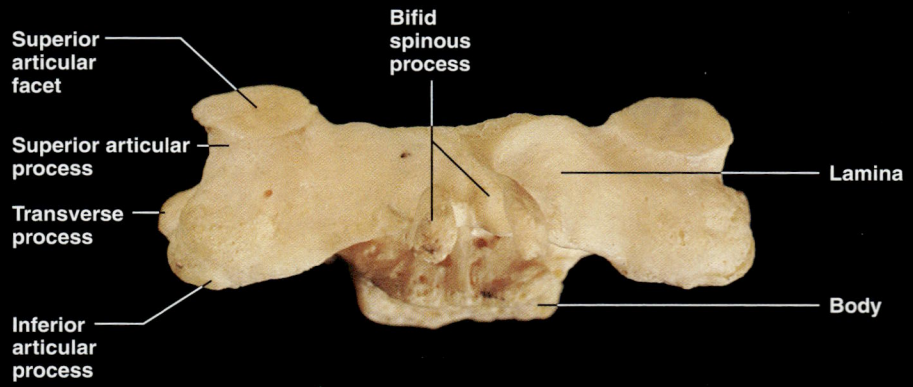

(c) fifth (typical) cervical vertebra, posterior view

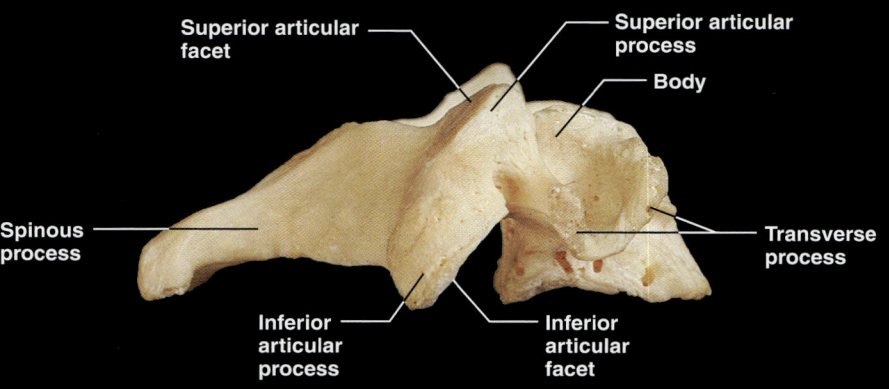

(d) fifth (typical) cervical vertebra, right lateral view

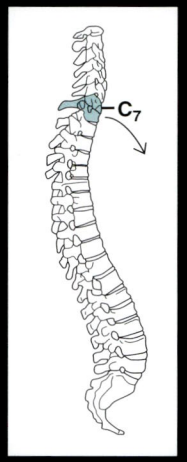

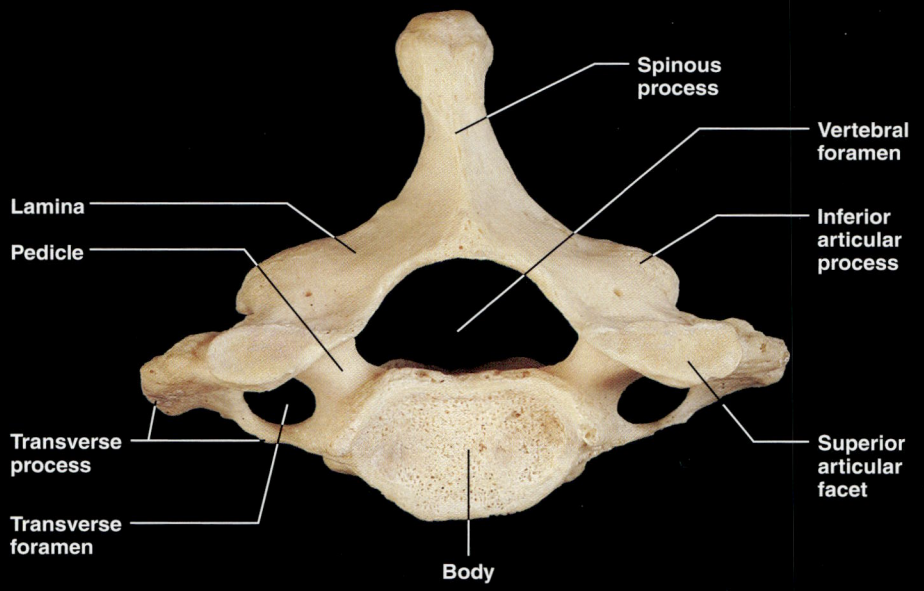

(e) vertebra prominens (C_7), superior view

Figure 19 Cervical vertebrae.

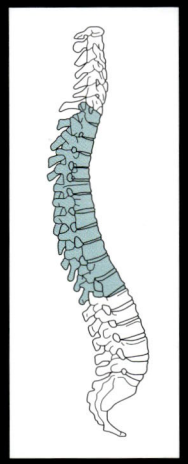

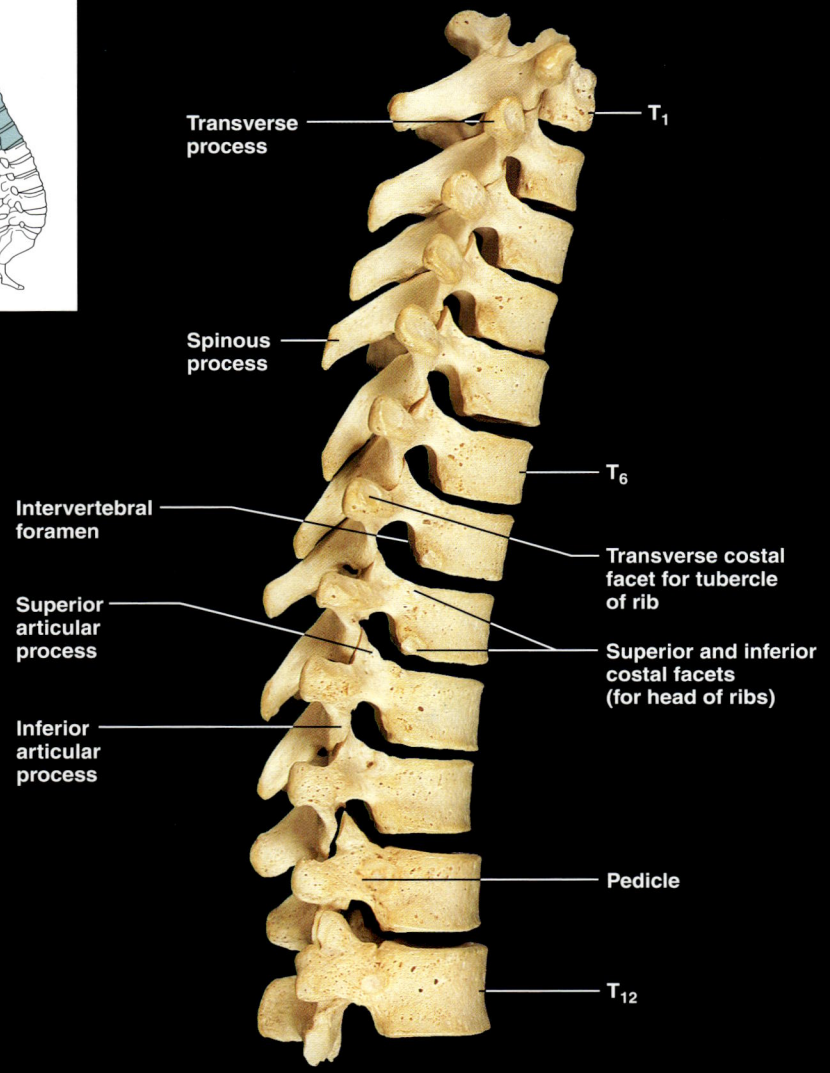

(a) articulated thoracic vertebrae, right lateral view

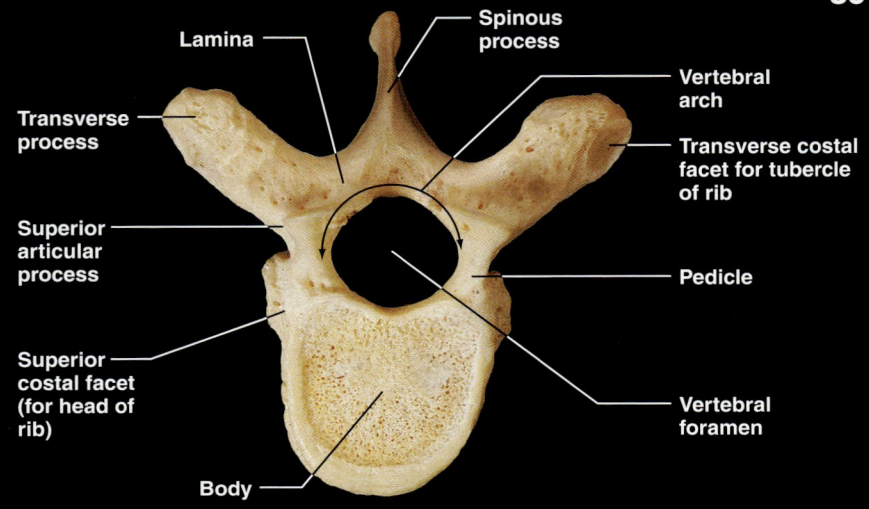

(b) seventh (typical) thoracic vertebra, superior view

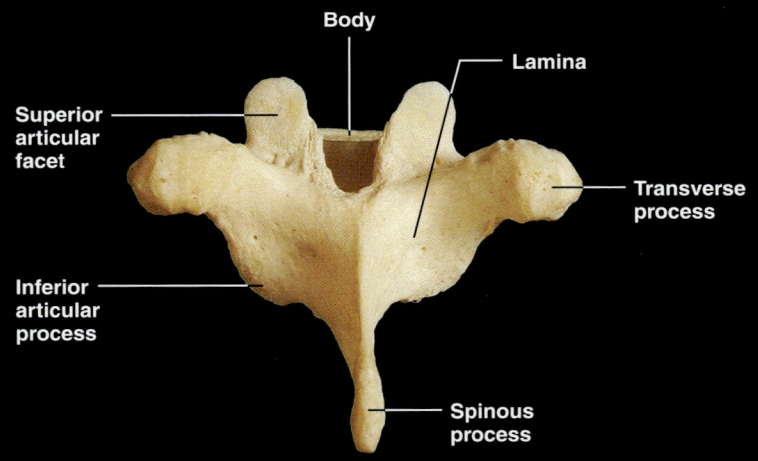

(c) seventh (typical) thoracic vertebra, posterior view

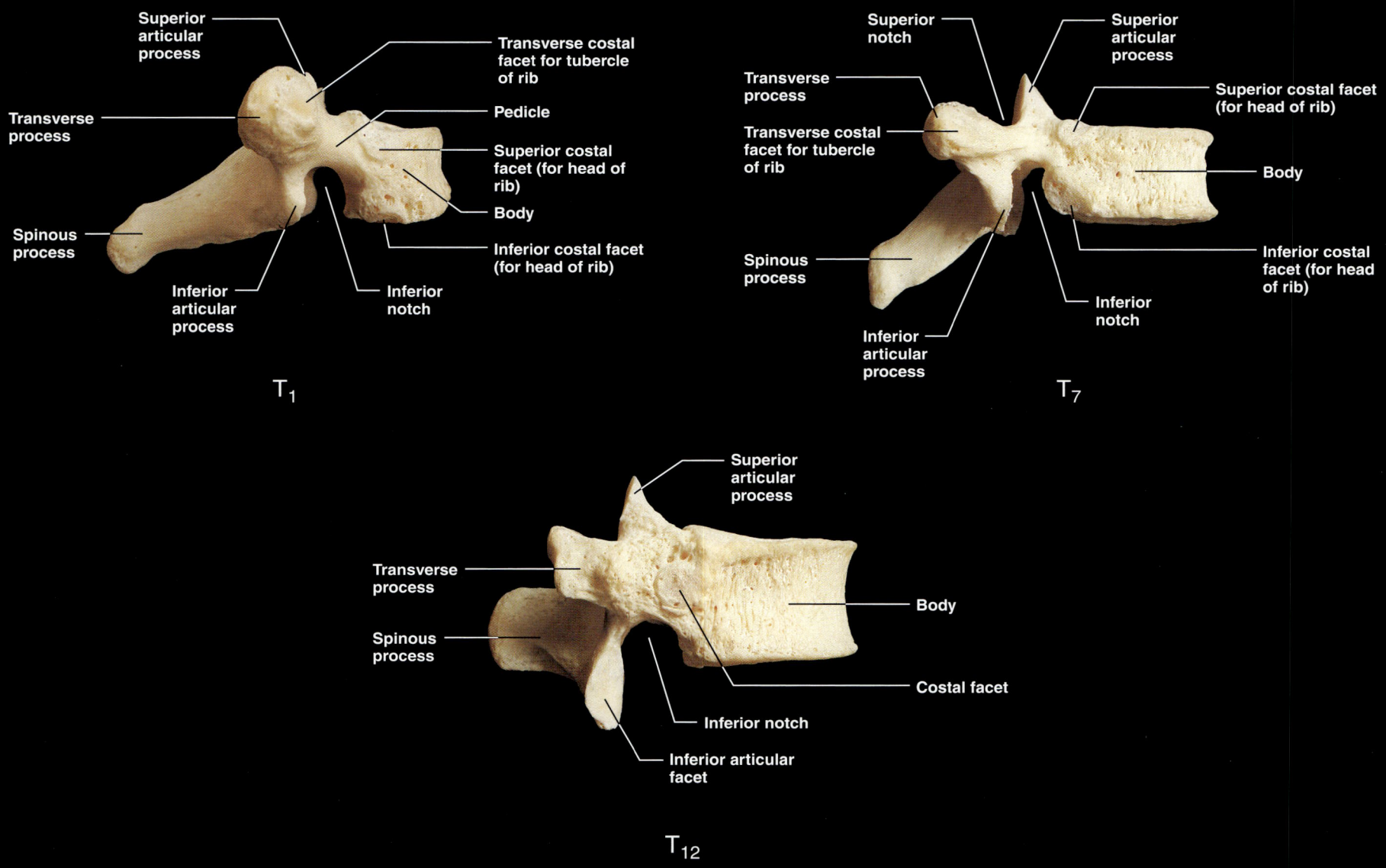

(d) comparison of T_1, T_7, and T_{12} in right lateral views

Figure 20 Thoracic vertebrae.

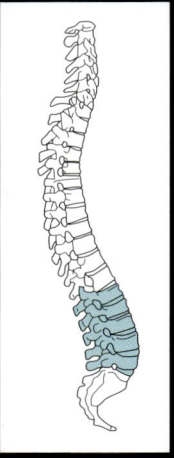

True ribs

False ribs

L1

L2

L3

L4

L5

Lumbar vertebrae

Intervertebral discs

Sacrum

Coccyx

(a) articulated lumbar vertebrae and rib cage, right lateral view

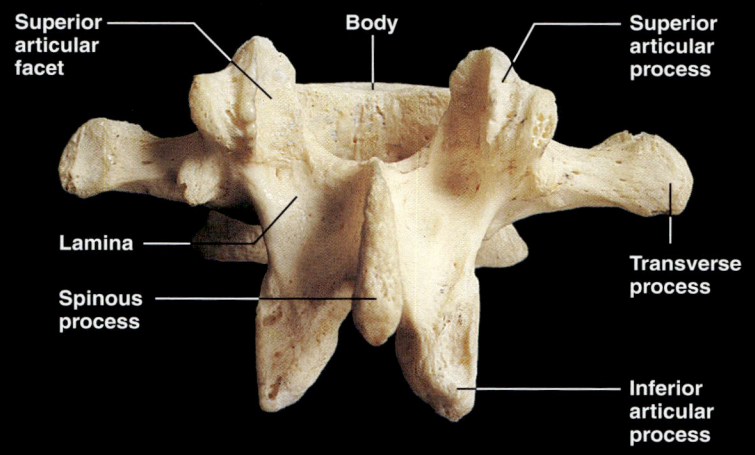

(c) second lumbar vertebra, posterior view

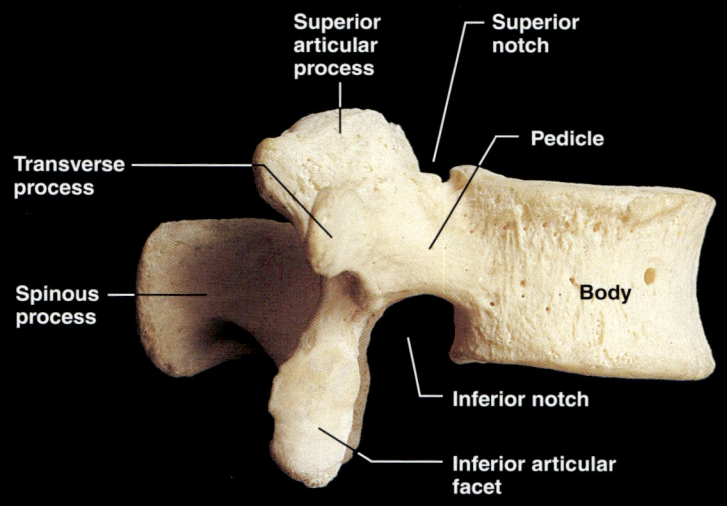

(d) second lumbar vertebra, right lateral view

Figure 21 Lumbar vertebrae.

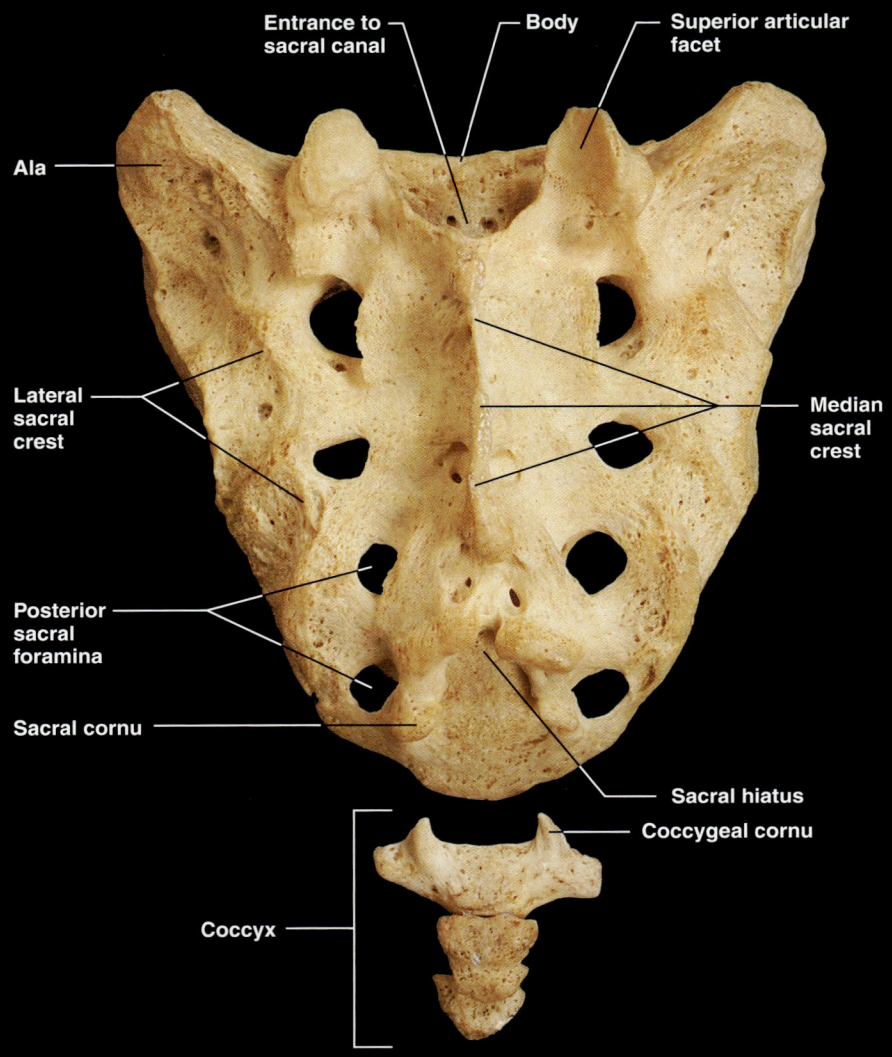

(a) posterior view

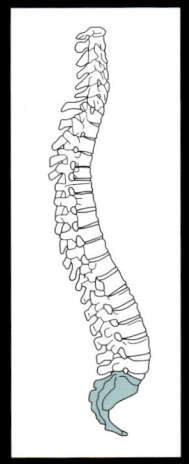

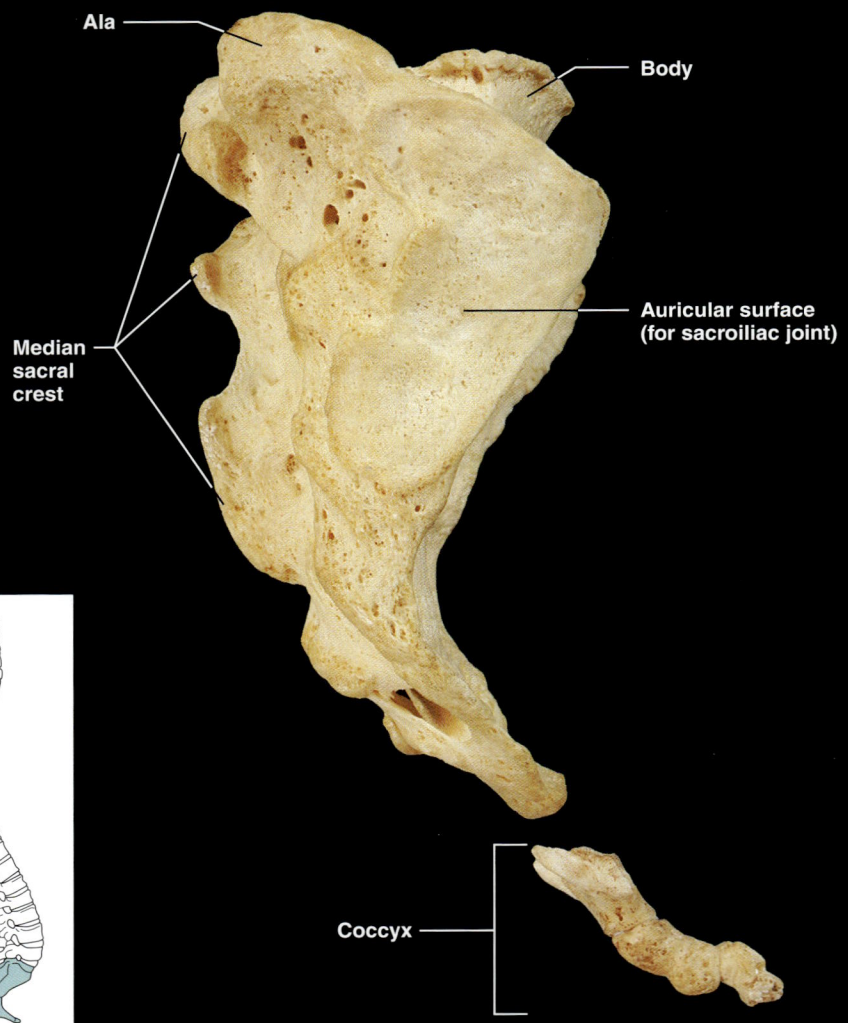

(b) right lateral view

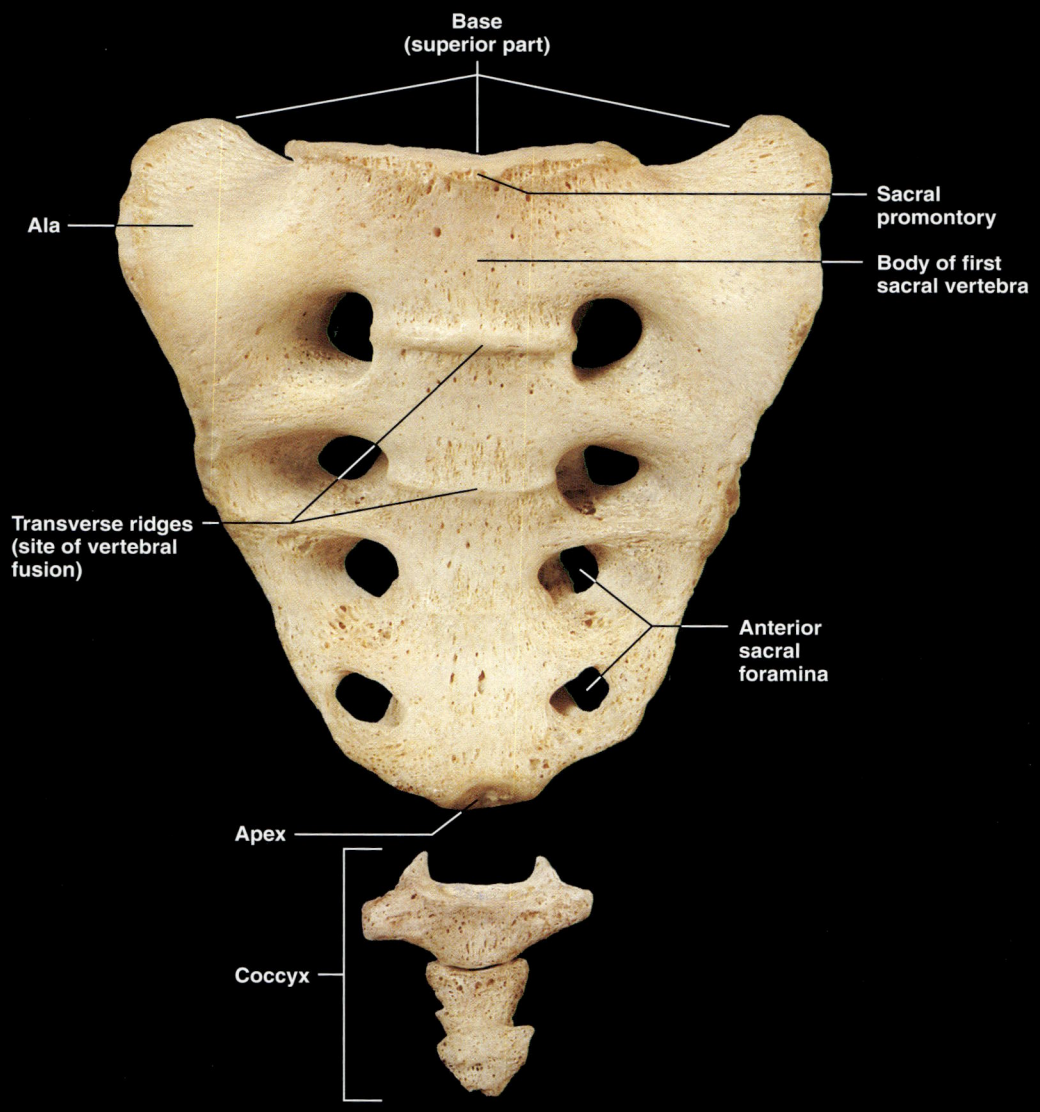

(c) anterior view

Figure 22 Sacrum and coccyx.

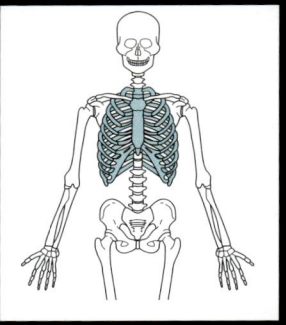

Jugular notch
Clavicular notch
Clavicle
Manubrium
Sternal angle
Body
Sternum
Xiphisternal joint
Xiphoid process
True ribs (1–7)
Intercostal spaces
Costal cartilage
False ribs (8–12)
Costal margin
L_1
Floating ribs (11, 12)

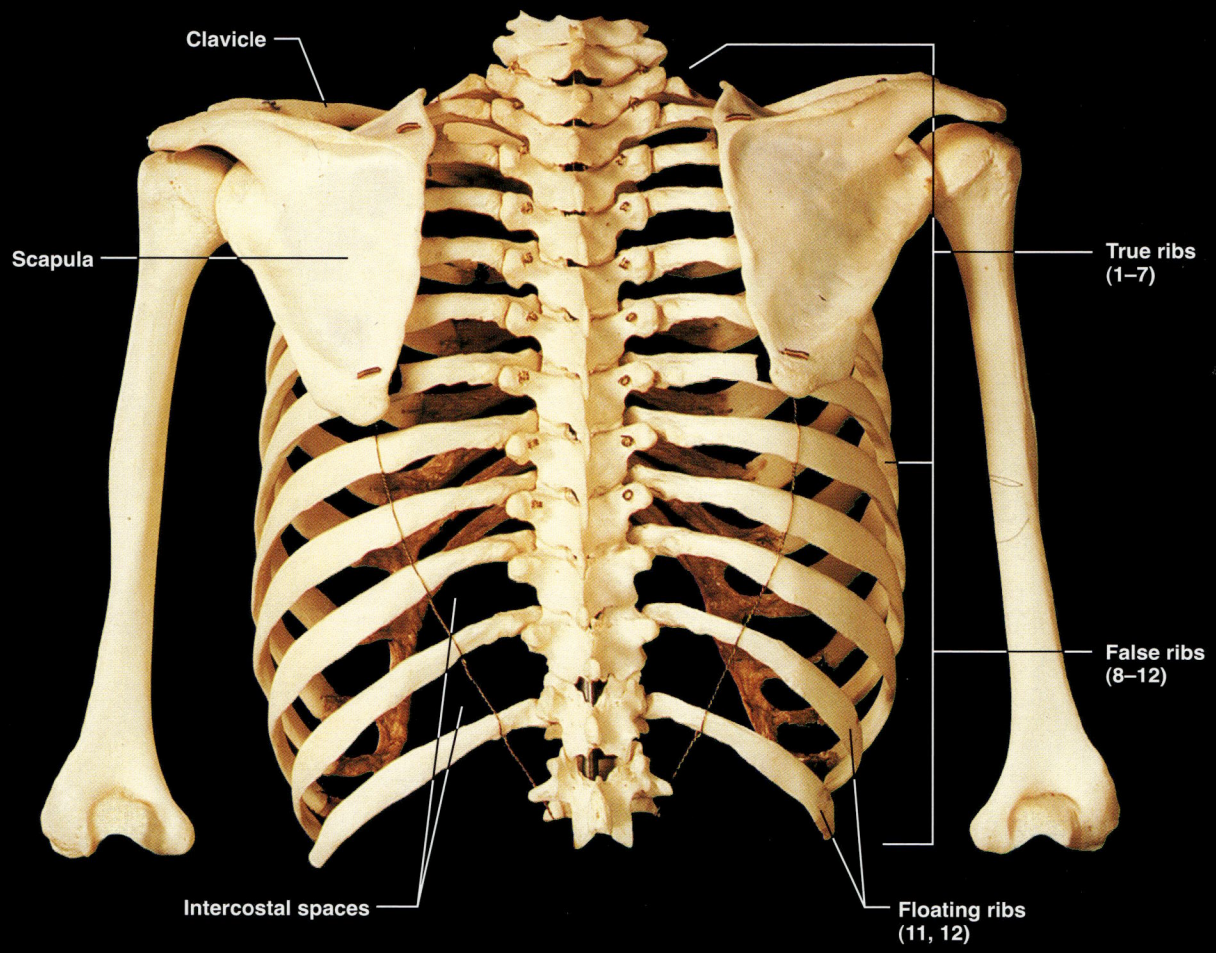

(b) posterior view

Figure 23 Thoracic cage

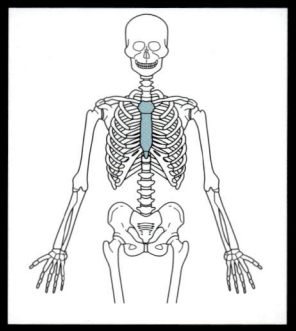

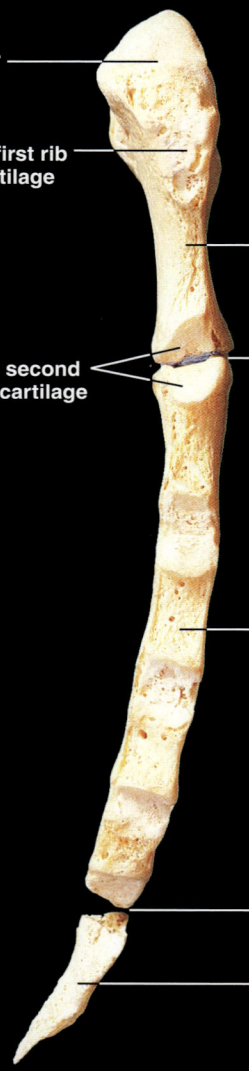

(c) sternum, right lateral view

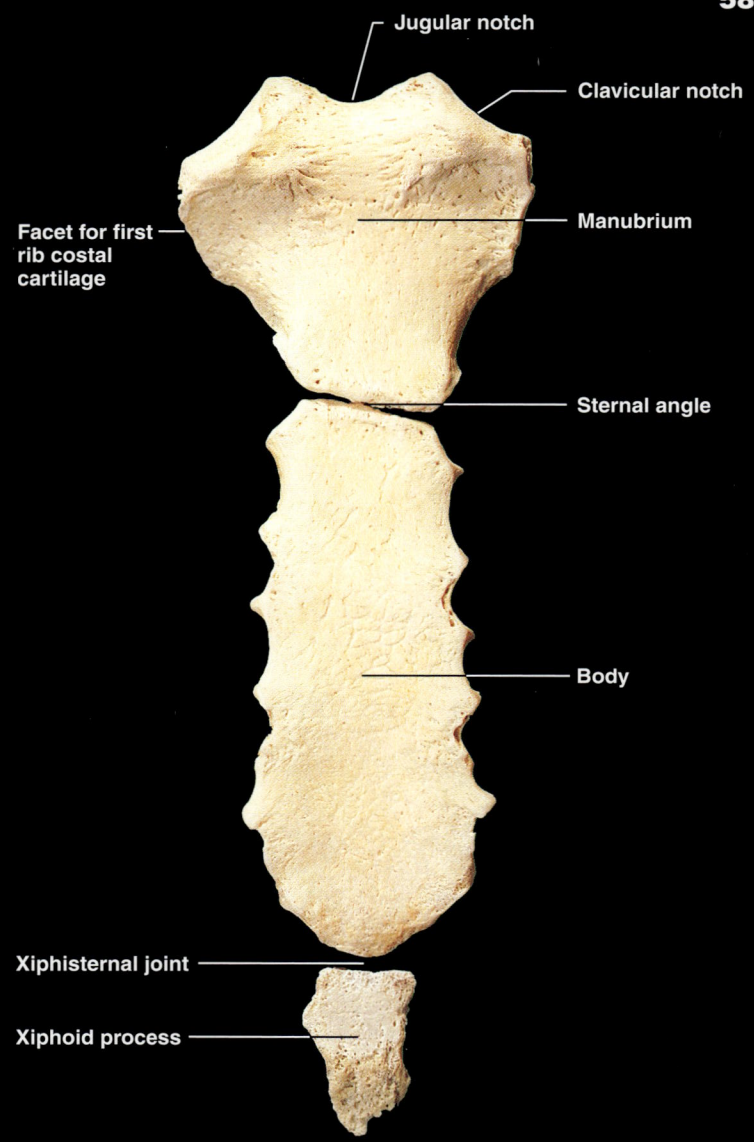

(d) sternum, anterior view

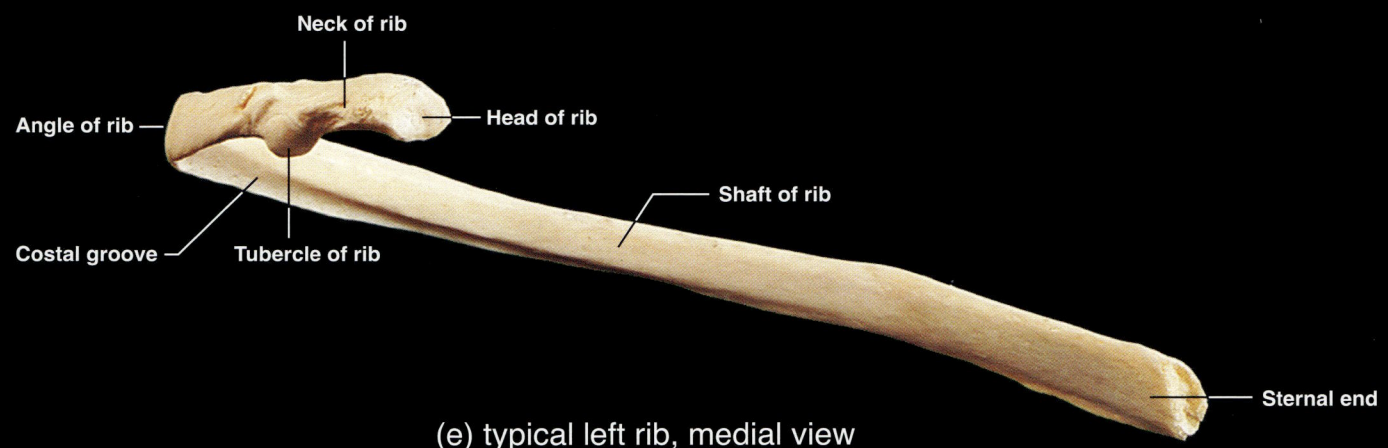

(e) typical left rib, medical view

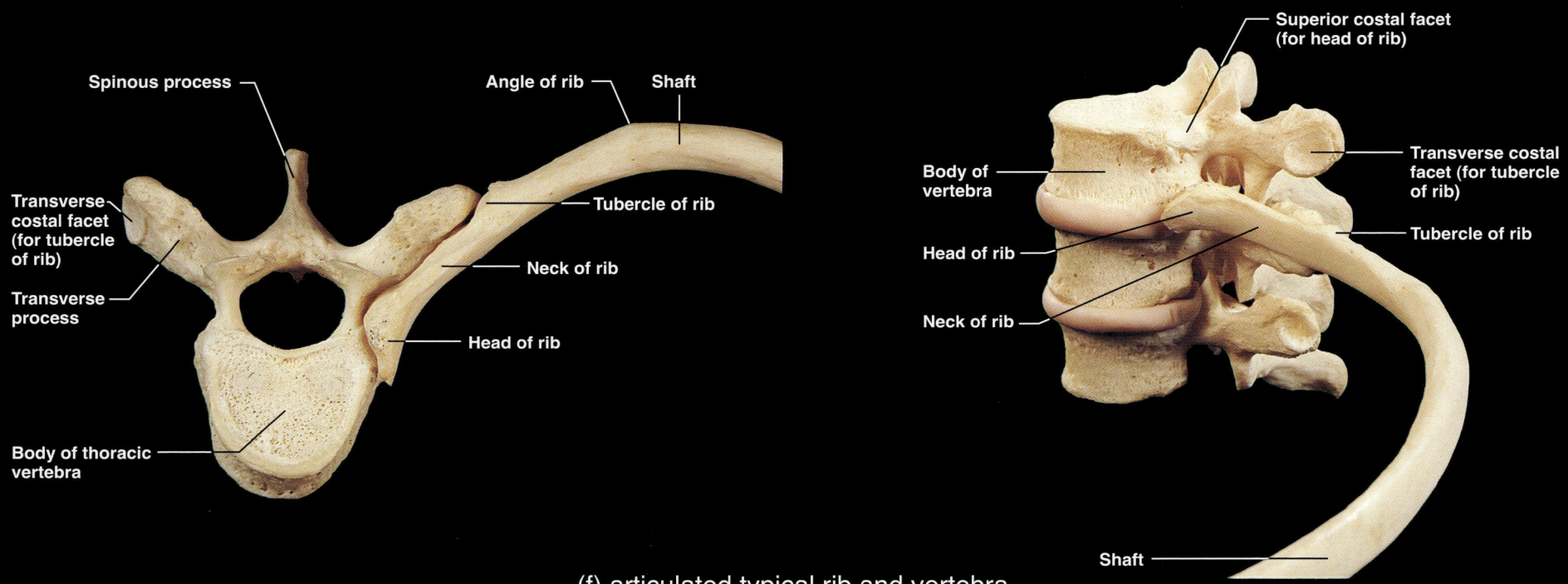

(f) articulated typical rib and vertebra, superior view (left); lateral view (right)

Figure 23 Thoracic cage (continued).

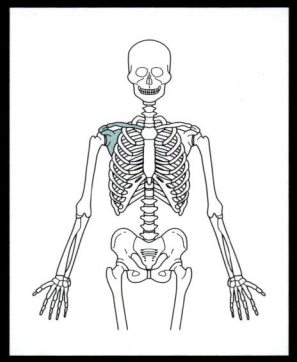

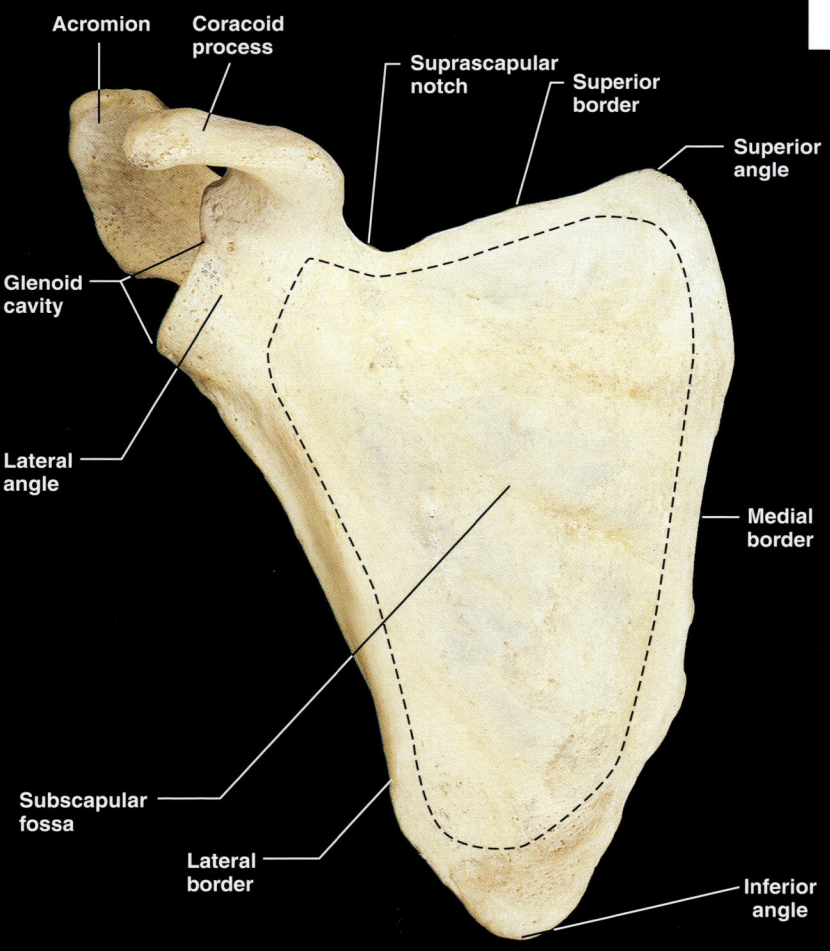

(a) right scapula, anterior view

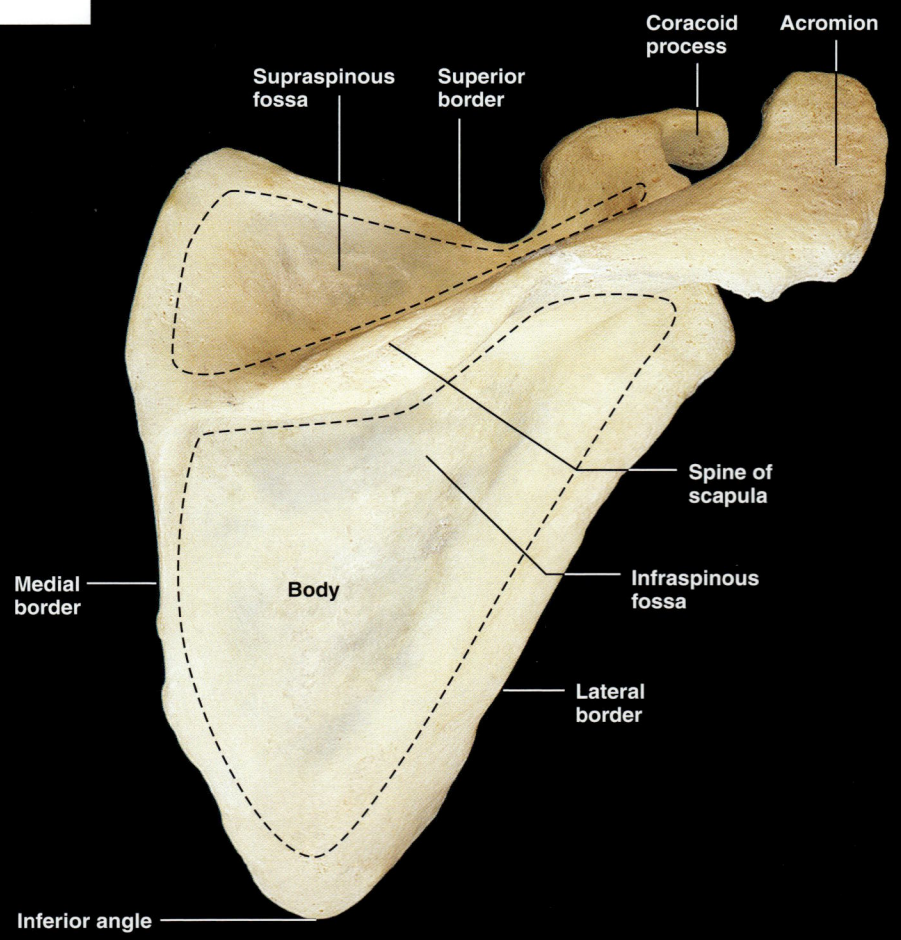

(b) right scapula, posterior view

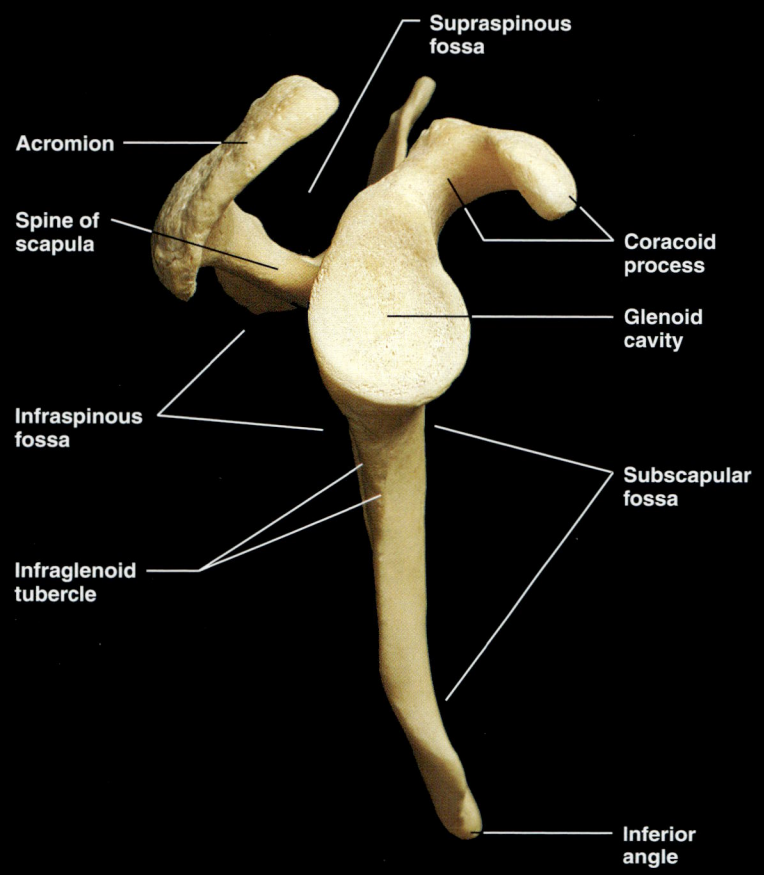

(c) right scapula, lateral aspect

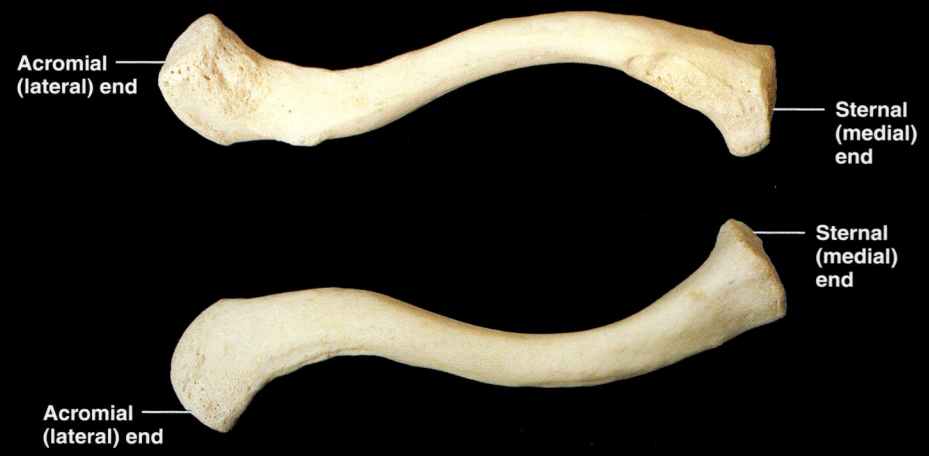

(d) right clavicle, inferior view (top) and superior view (bottom)

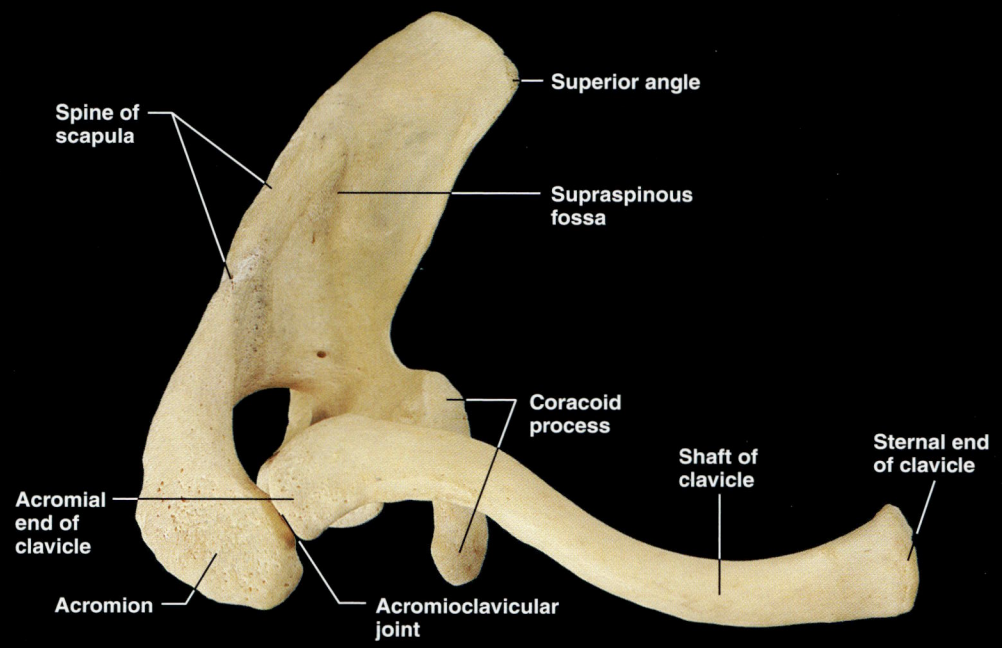

(e) articulated right clavicle and scapula, superior view

Figure 24 Scapula and clavicle.

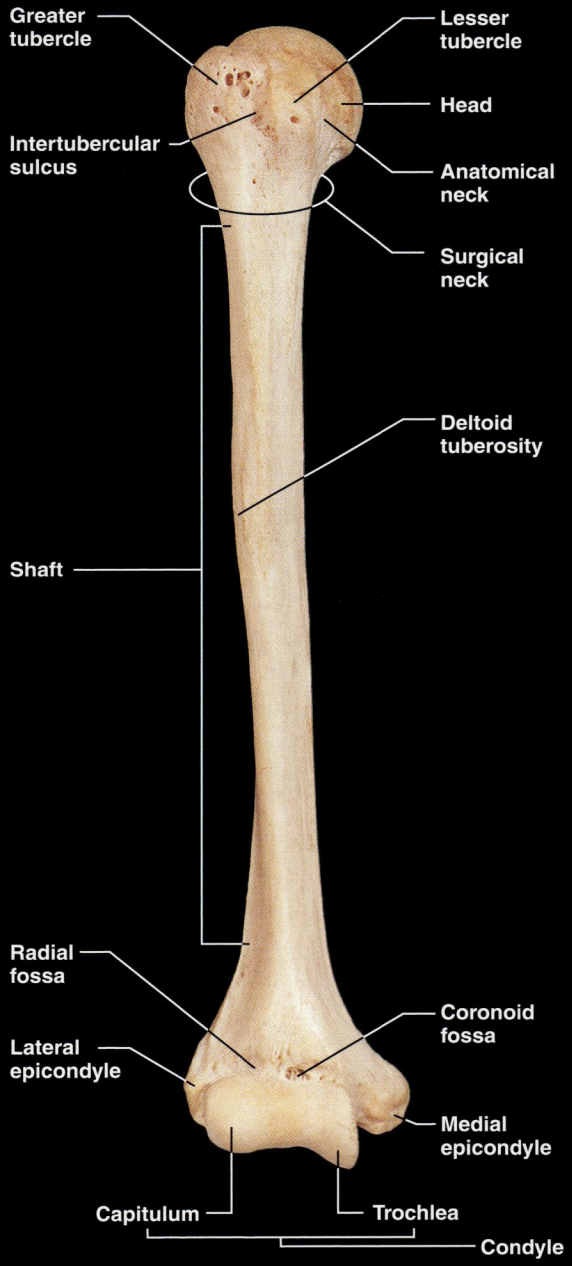

(a) anterior view

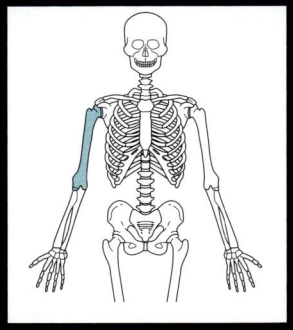

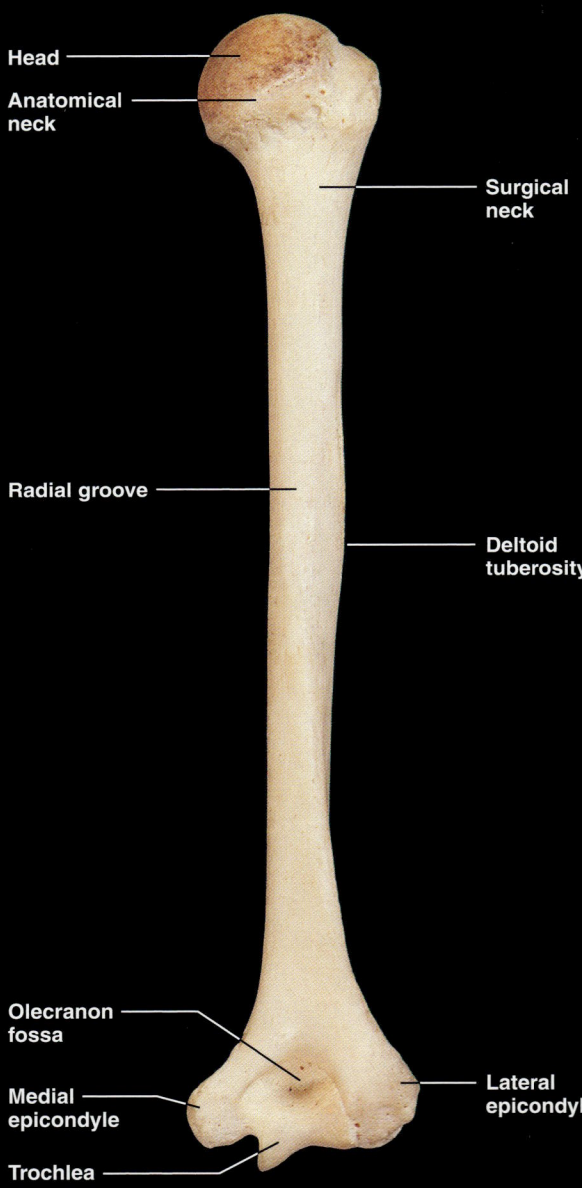

(b) posterior view

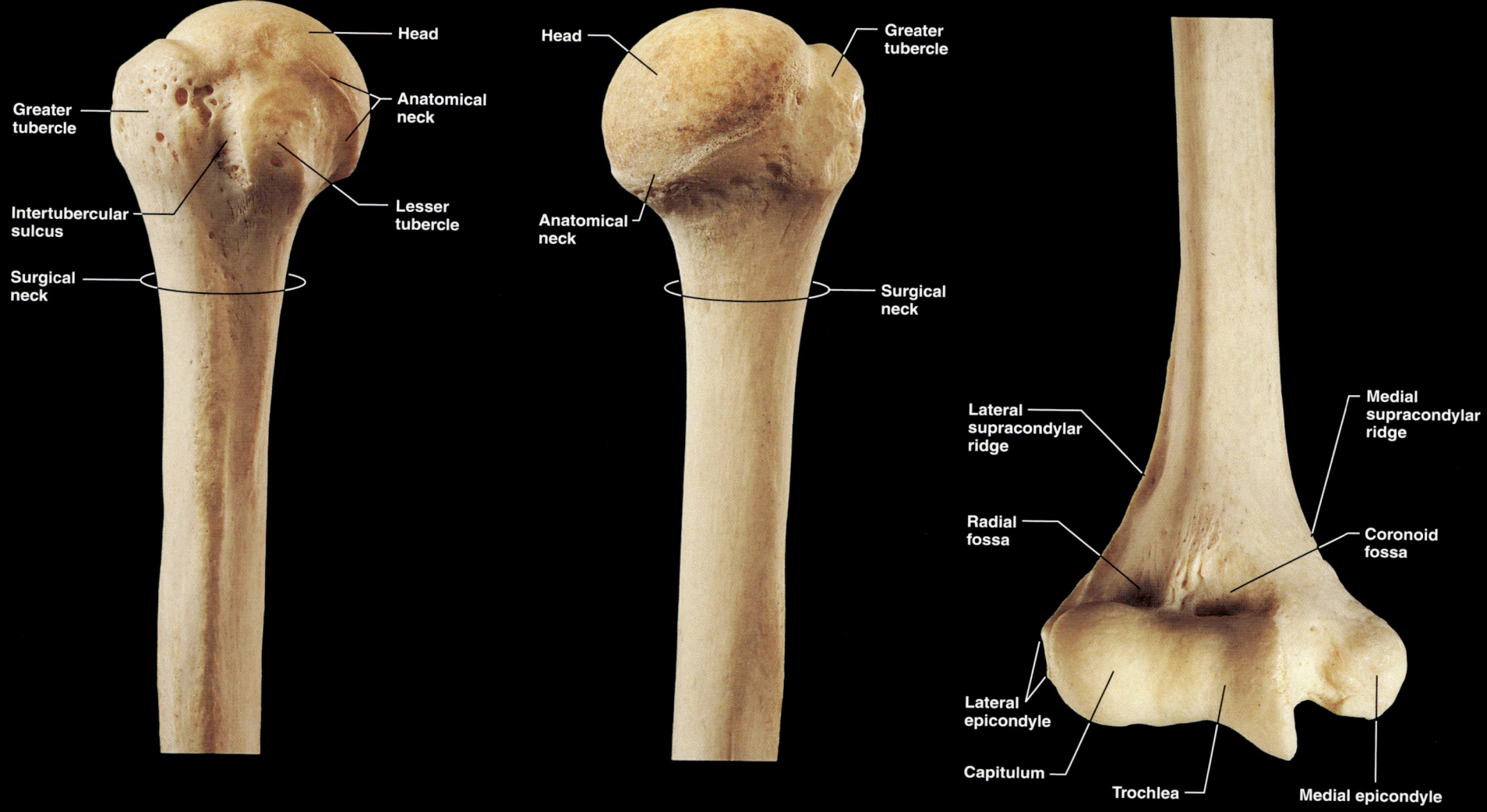

(c) proximal end, anterior view

(d) proximal end, posterior view

(e) distal end, anterior view

Figure 25 Right humerus.

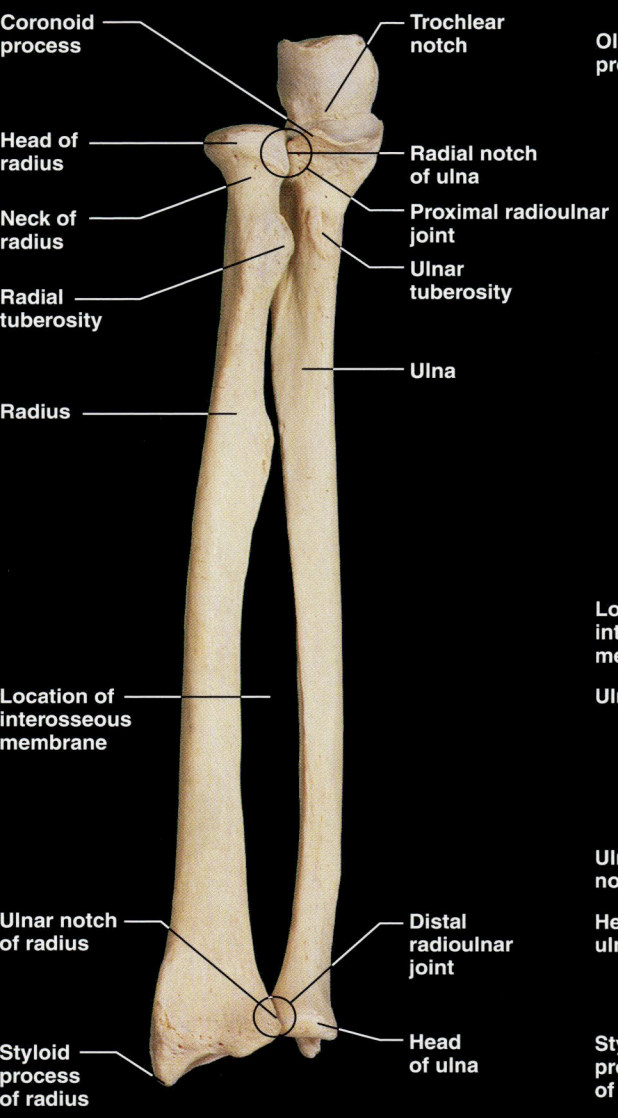

(a) articulated right ulna and radius, anterior view

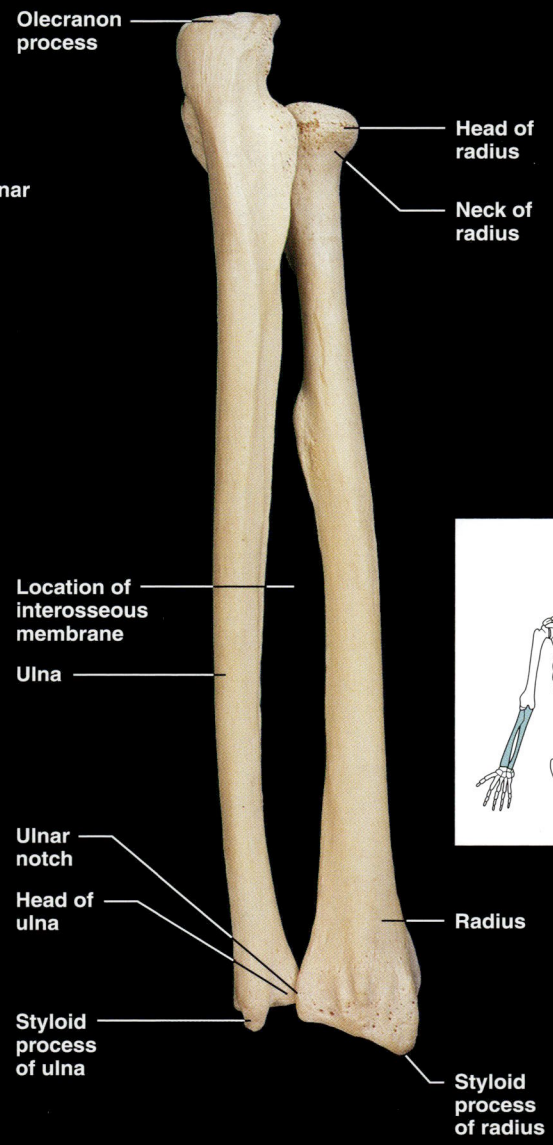

(b) articulated right ulna and radius, posterior view

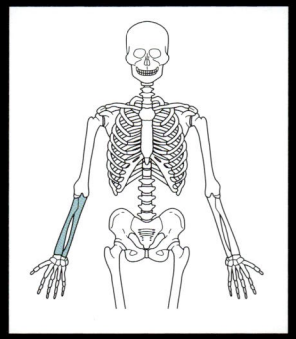

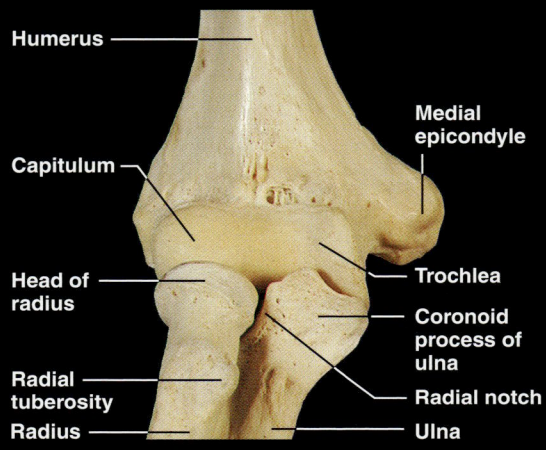

(c) articulated right humerus, ulna, and radius, anterior view

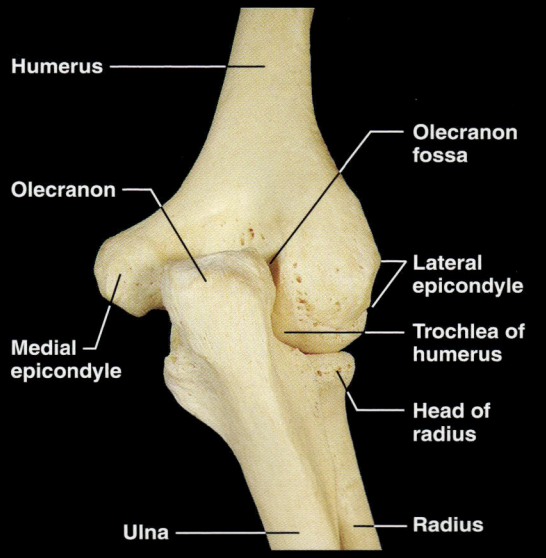

(d) articulated right humerus, ulna, and radius, posterior view

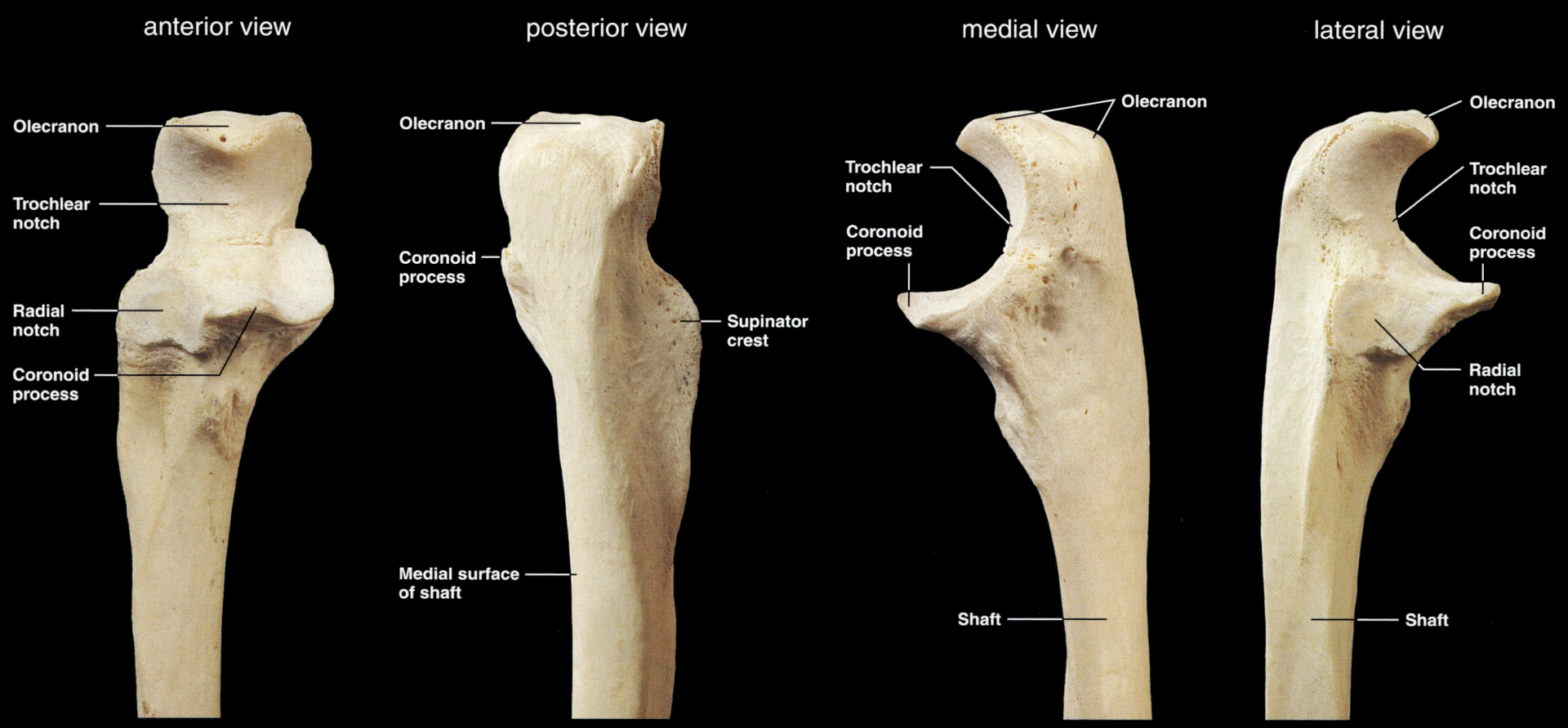

(e) right ulna, proximal end

Figure 26 Right ulna and radius.

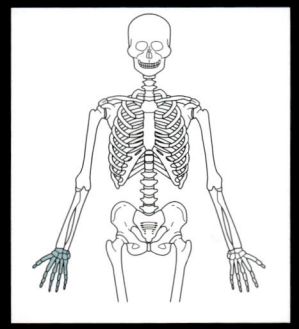

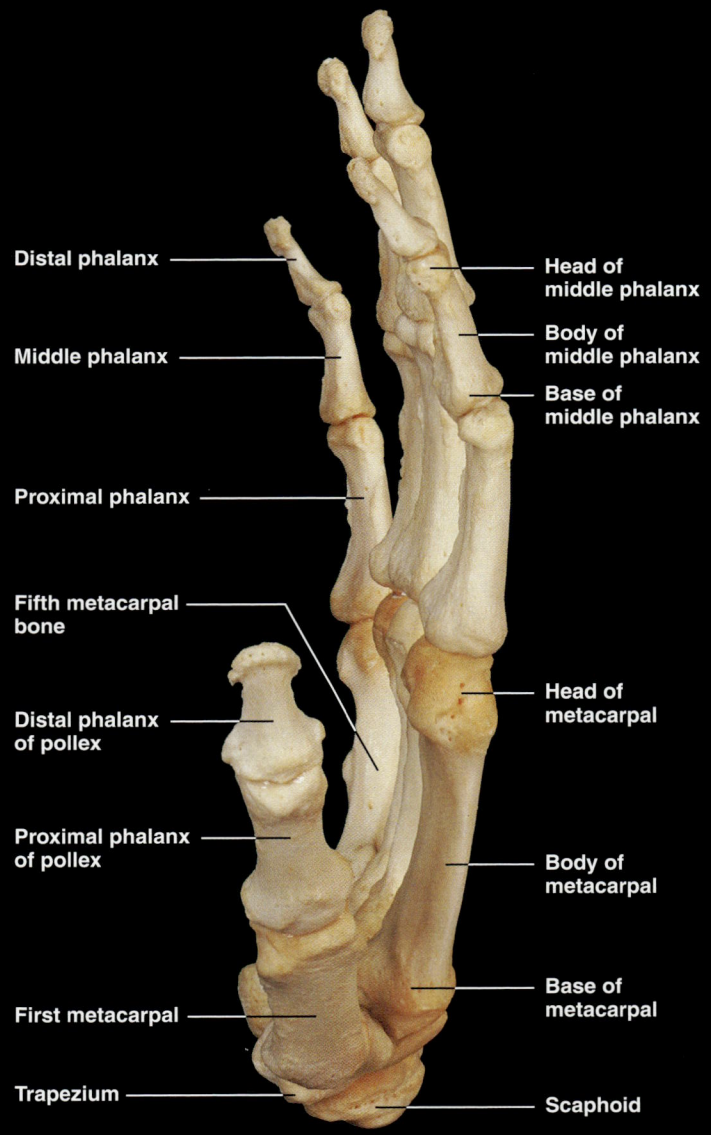

(a) lateral aspect

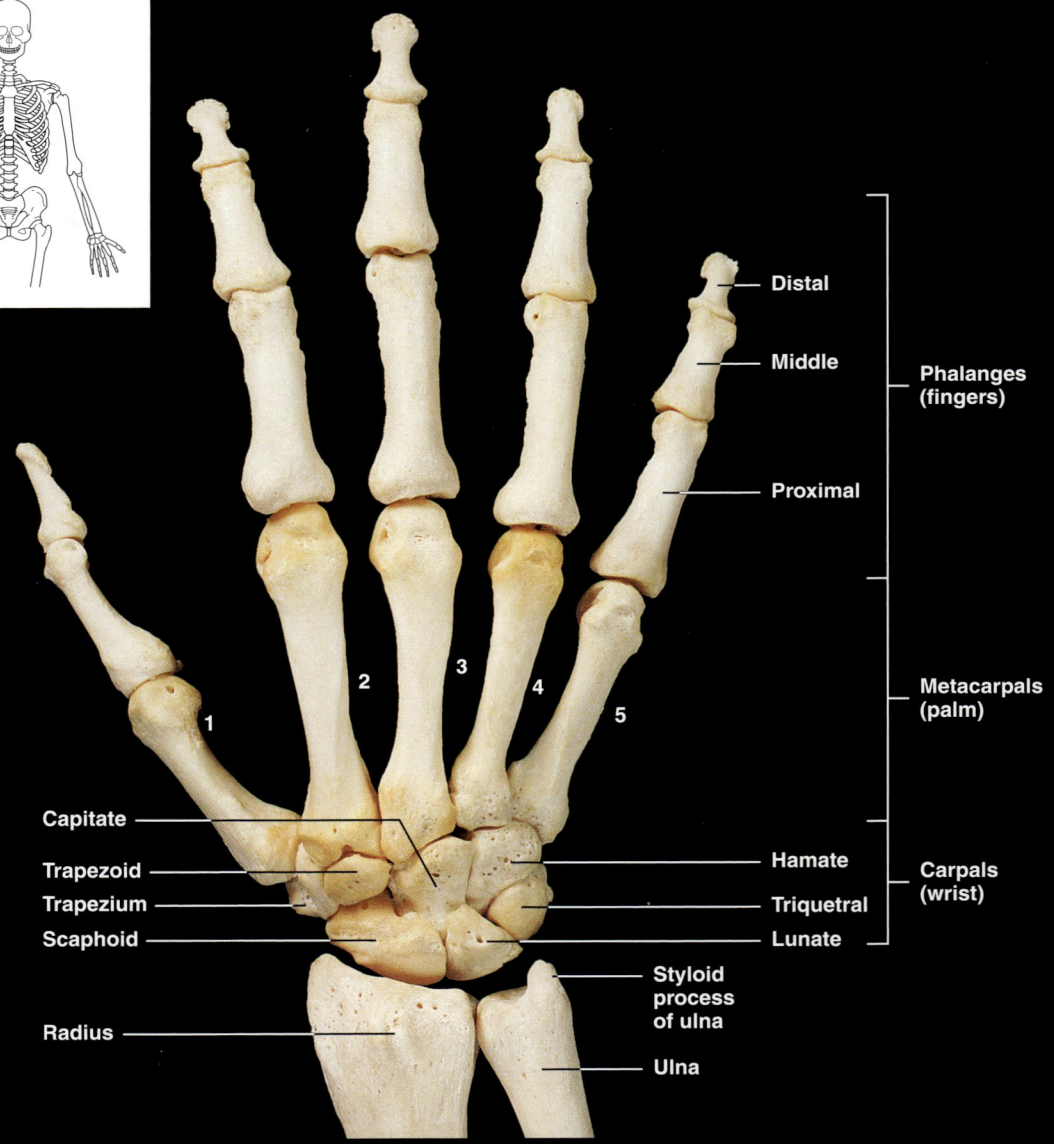

(b) dorsal aspect

Figure 27 Bones of the right hand.

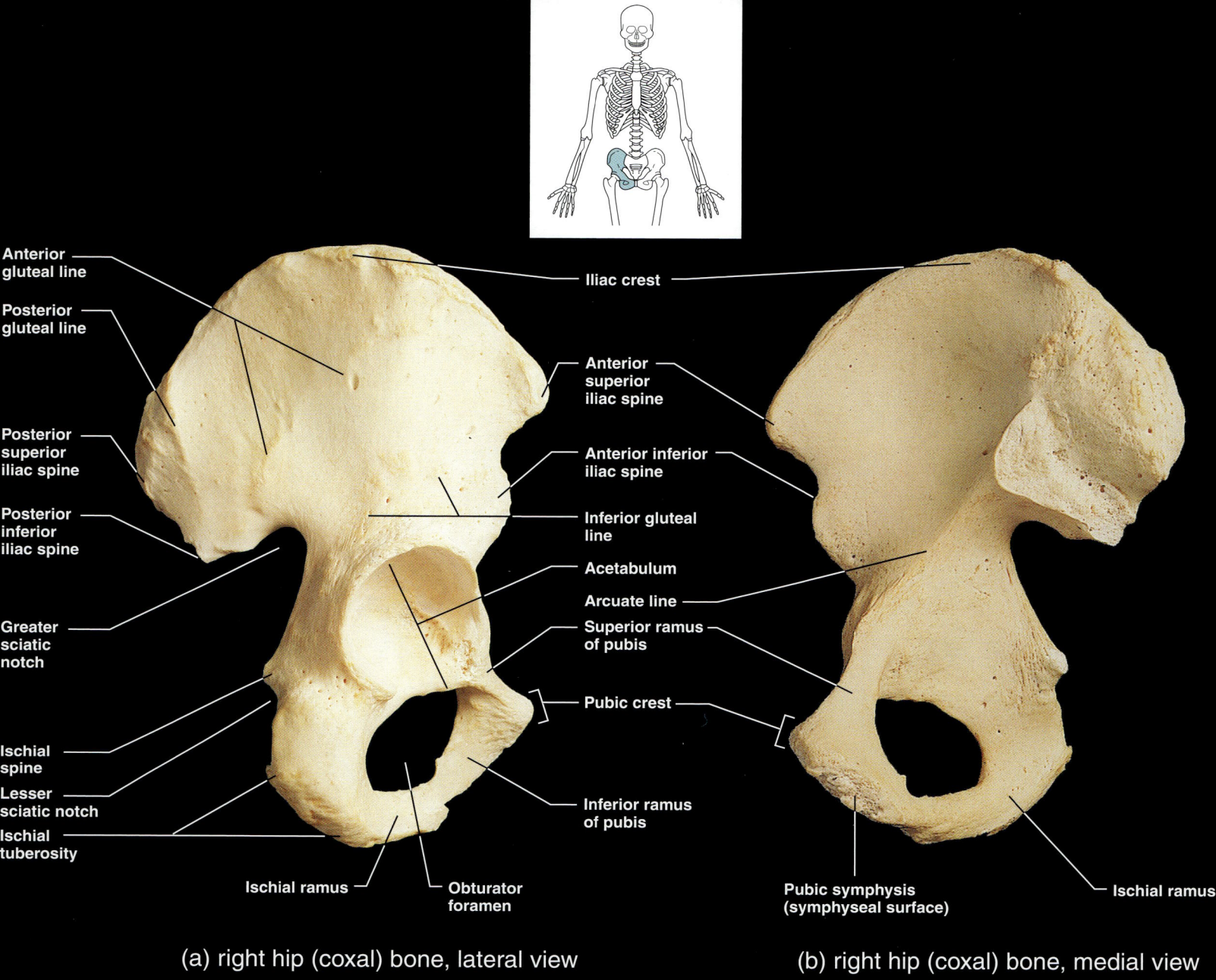

(a) right hip (coxal) bone, lateral view

(b) right hip (coxal) bone, medial view

Figure 28 Bones of the male pelvis.

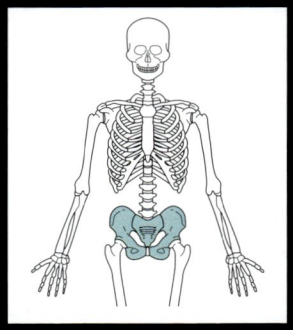

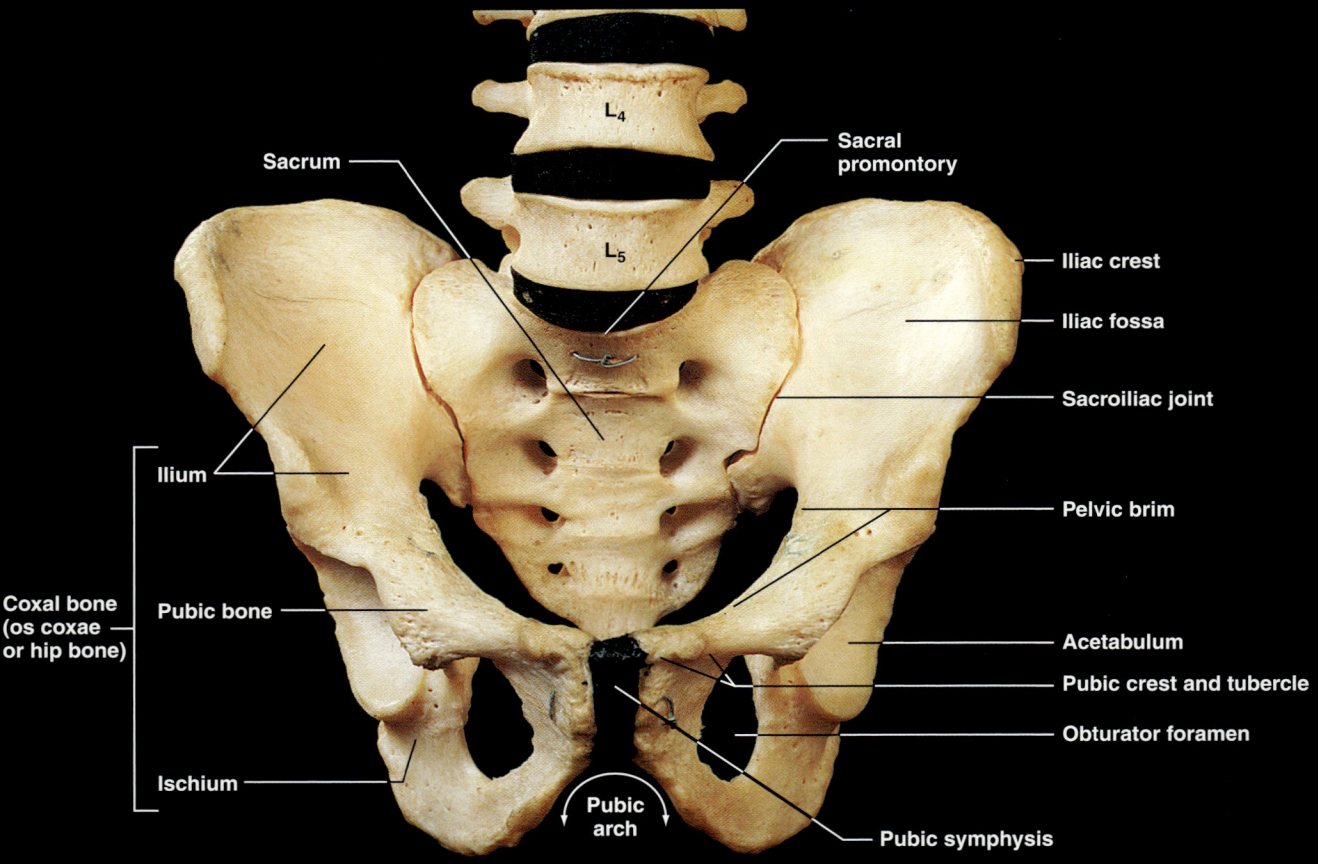

(c) articulated male pelvis, anterior view

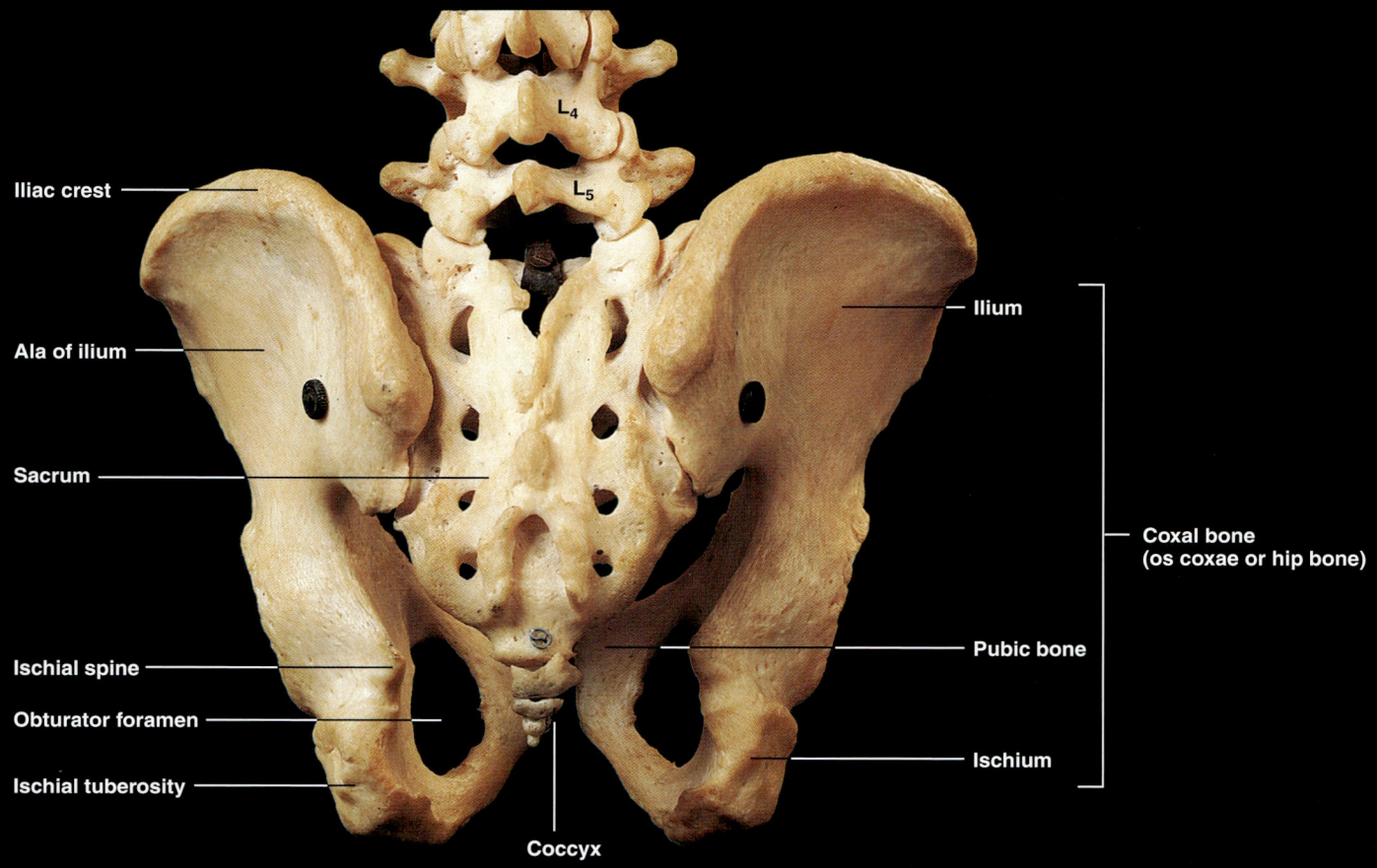

(d) articulated male pelvis, posterior view

Figure 28 Bones of the male pelvis (continued).

Head
Intertrochanteric line
Lesser trochanter
Adductor tubercle
Medial epicondyle
Medial condyle

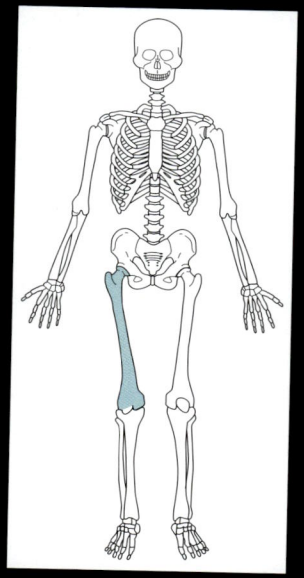

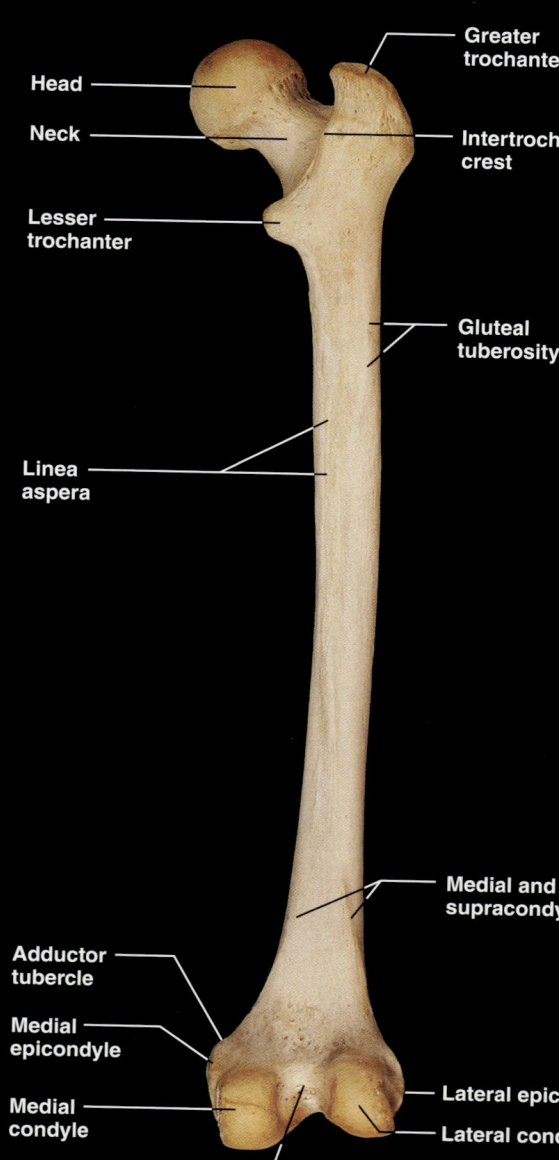
Head
Neck
Lesser trochanter
Linea aspera
Adductor tubercle
Medial epicondyle
Medial condyle
Greater trochante
Intertroch crest
Gluteal tuberosity
Medial and supracondy
Lateral epic
Lateral con

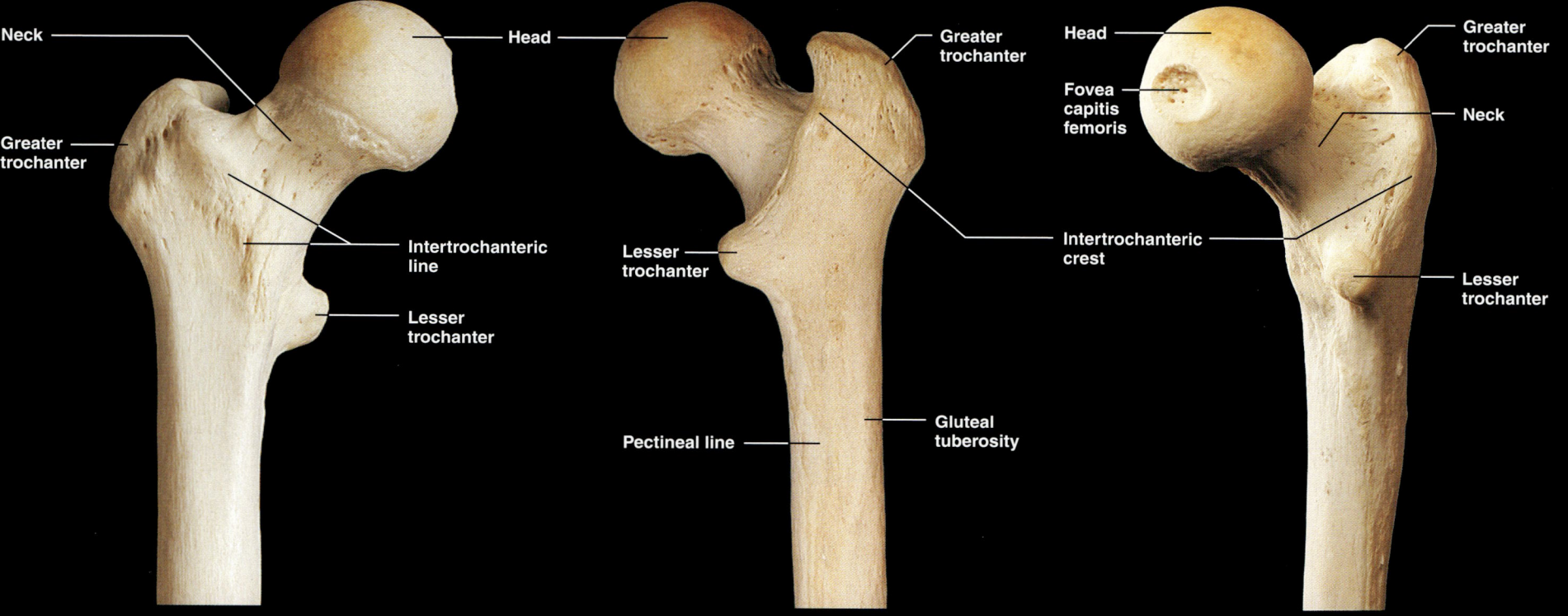

(c) proximal end, anterior view

(d) proximal end, posterior view

(e) proximal end, medial view

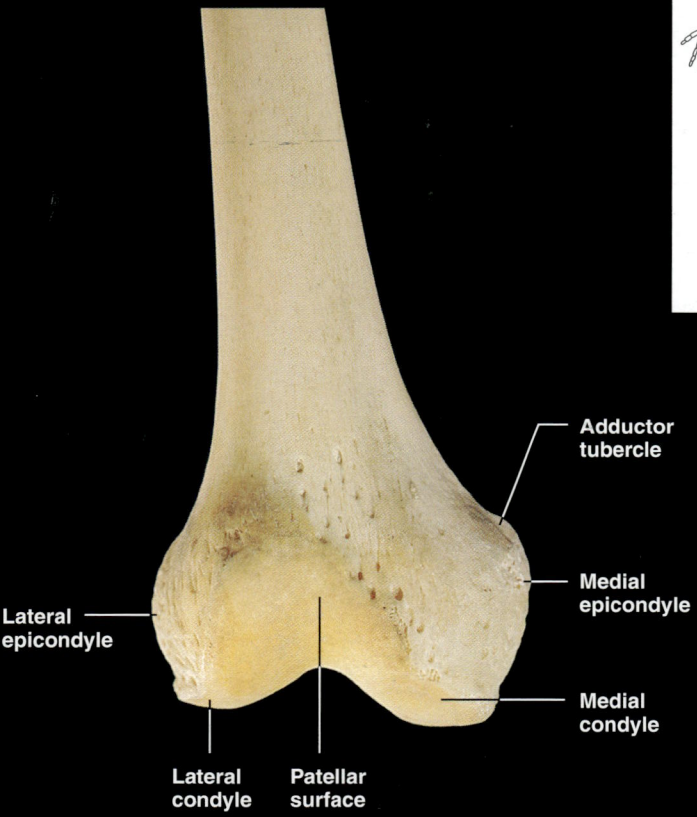

(f) distal end, anterior view

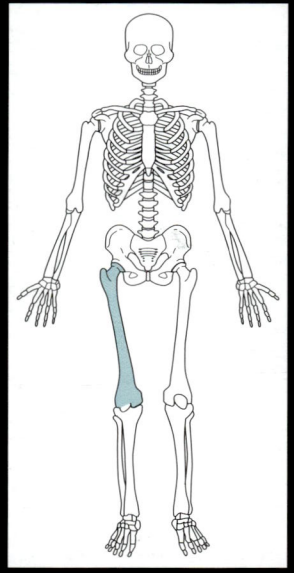

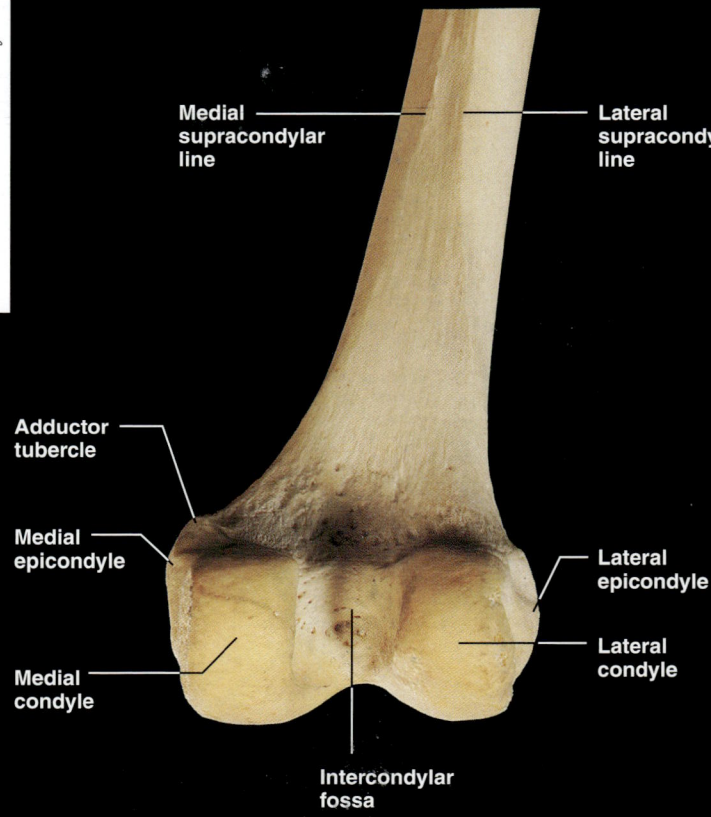

(g) distal end, posterior view

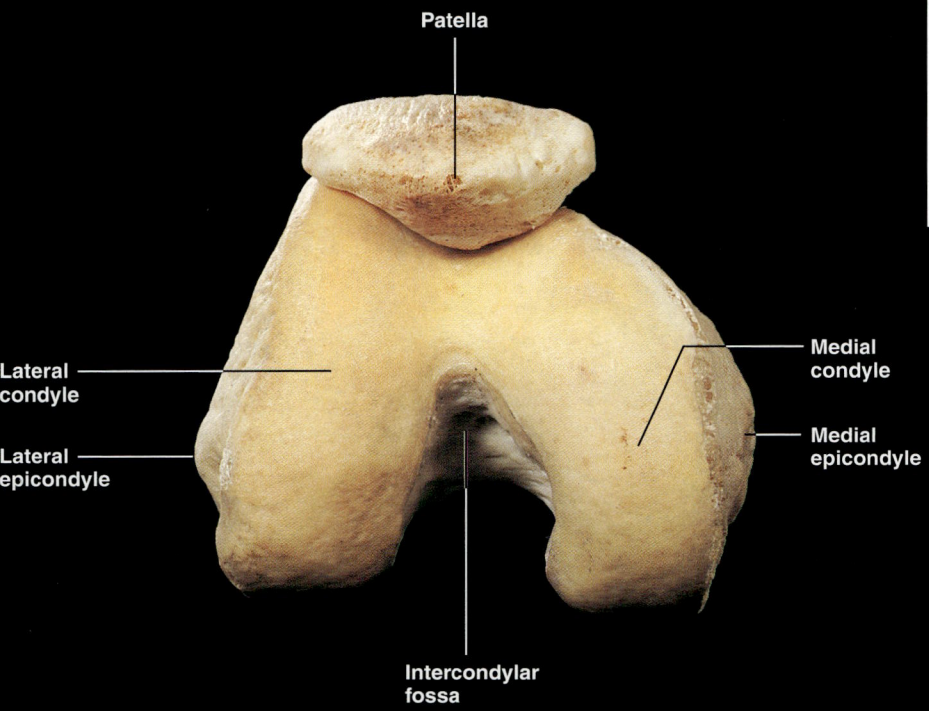

(h) Articulated right femur and patella, inferior view with knee extended

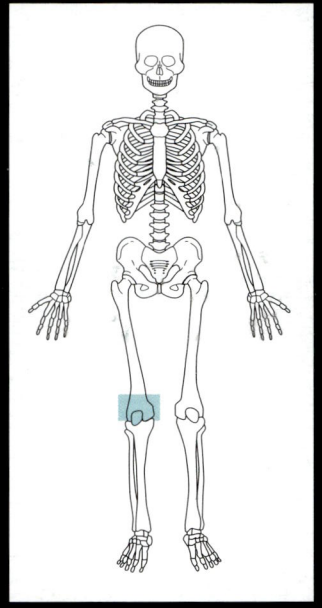

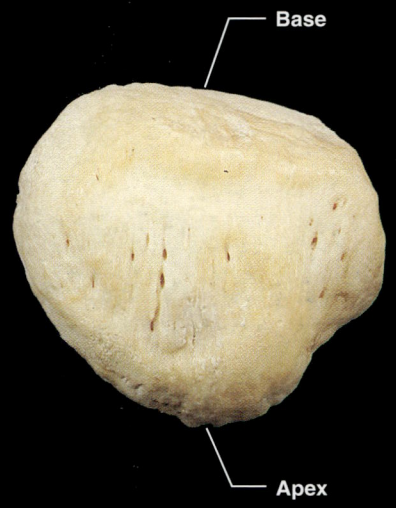

(i) Right patella, anterior surface

Figure 29 Right femur (continued).

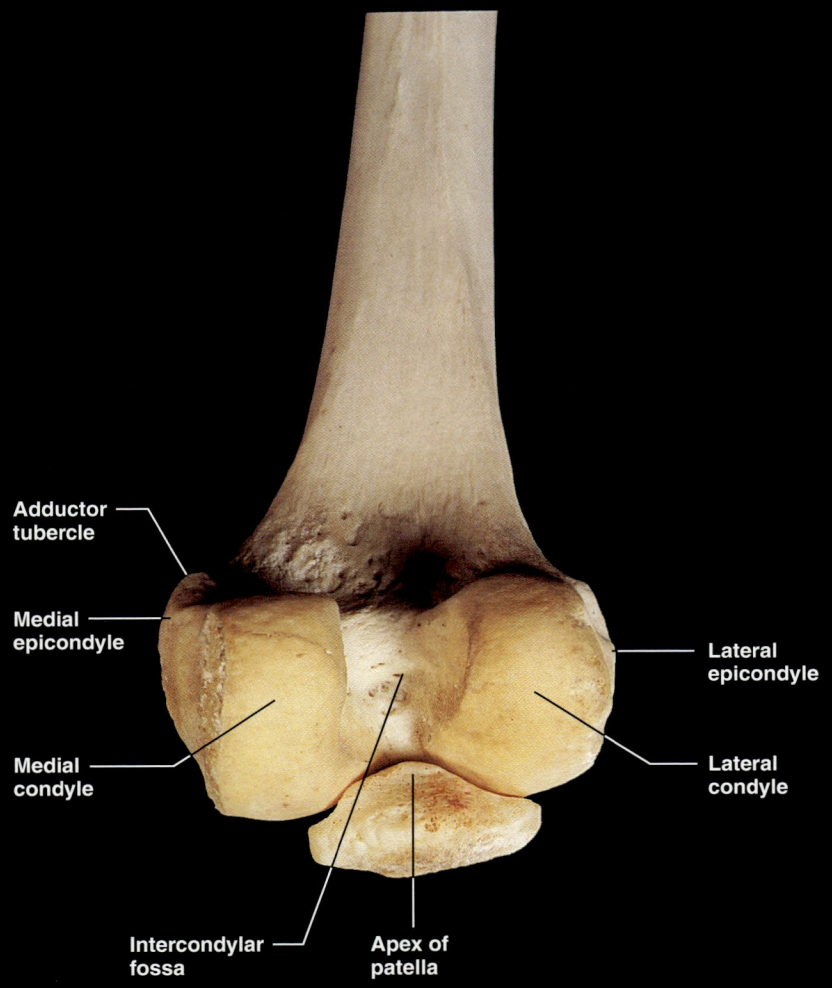

(j) Articulated right femur and patella, inferior posterior view with knee flexed

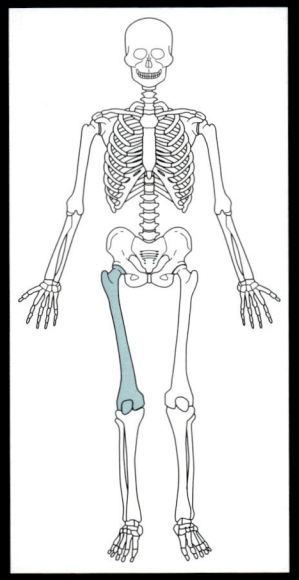

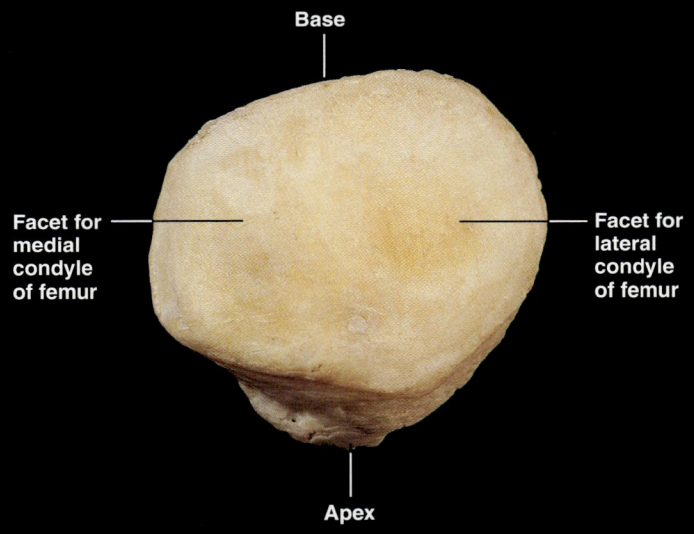

(k) Right patella, posterior surface

Figure 29 Right femur (continued).

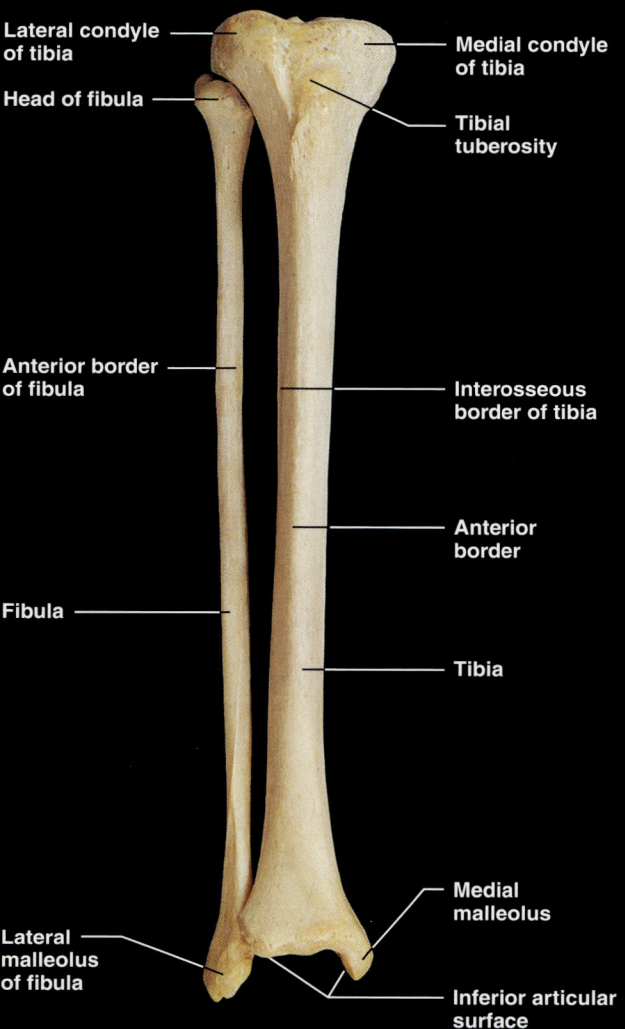
Lateral condyle of tibia
Head of fibula
Medial condyle of tibia
Tibial tuberosity
Anterior border of fibula
Interosseous border of tibia
Anterior border
Fibula
Tibia
Medial malleolus
Lateral malleolus of fibula
Inferior articular surface

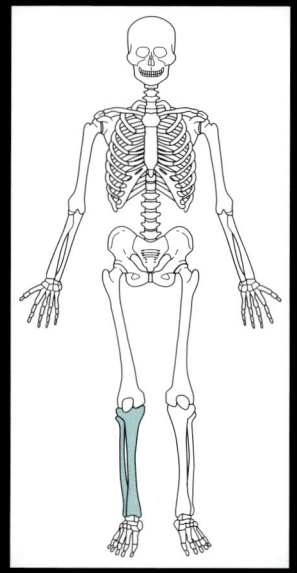

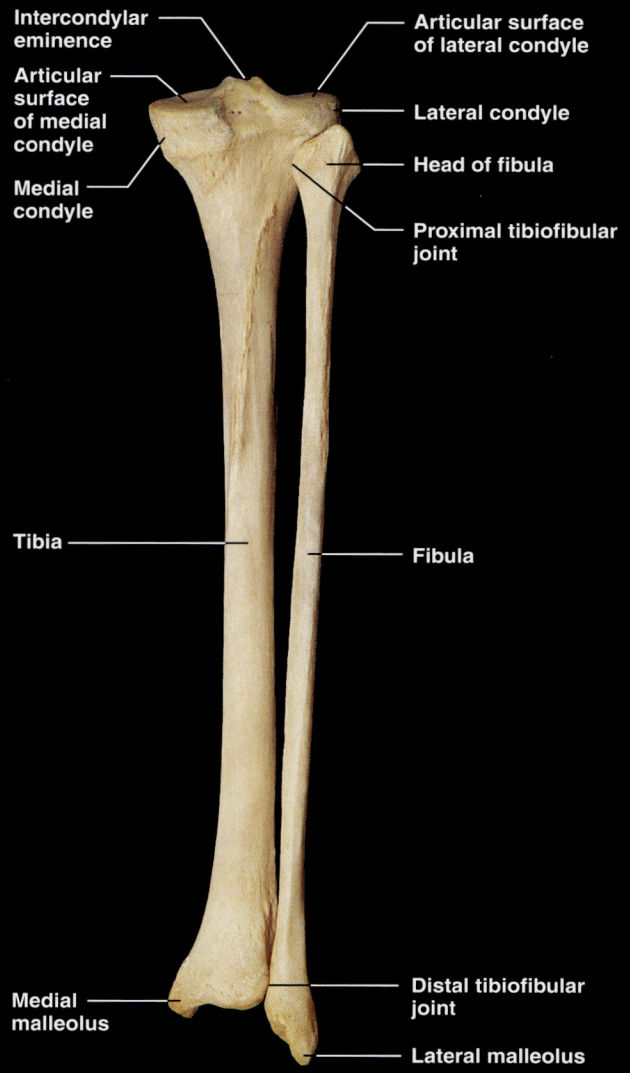
Intercondylar eminence
Articular surface of medial condyle
Medial condyle
Articular surface of lateral condyle
Lateral condyle
Head of fibula
Proximal tibiofibular joint
Tibia
Fibula
Medial malleolus
Distal tibiofibular joint
Lateral malleolus

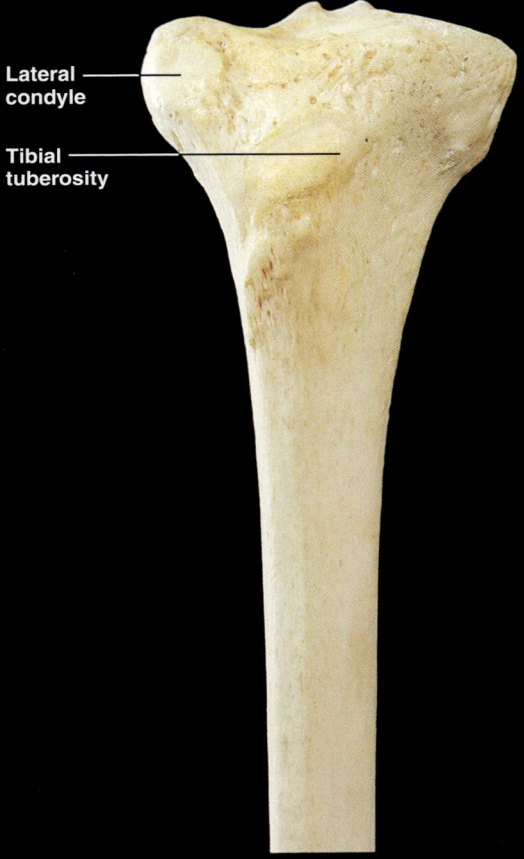

(c) right tibia, proximal end, anterior view

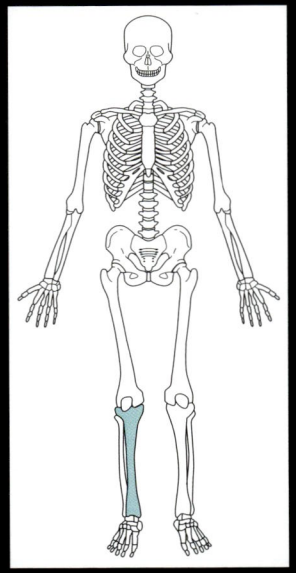

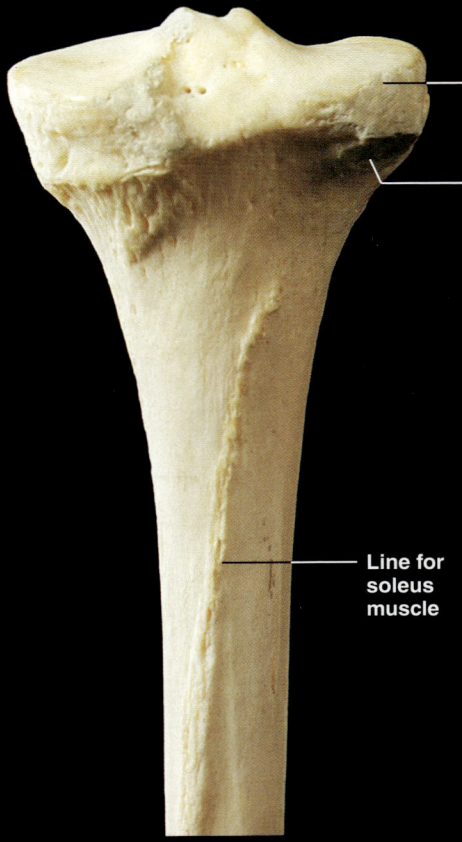

(d) right tibia, proximal end, posterior view

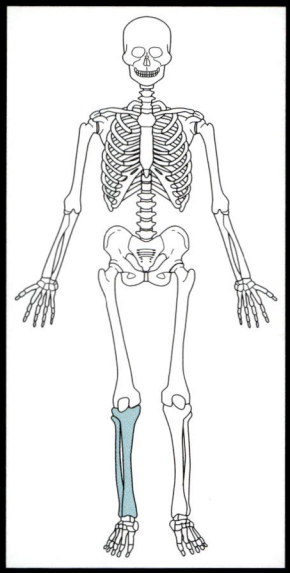

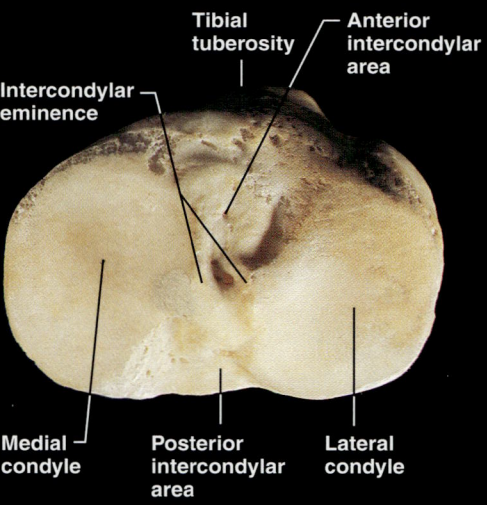

(e) right tibia, proximal end, articular surface

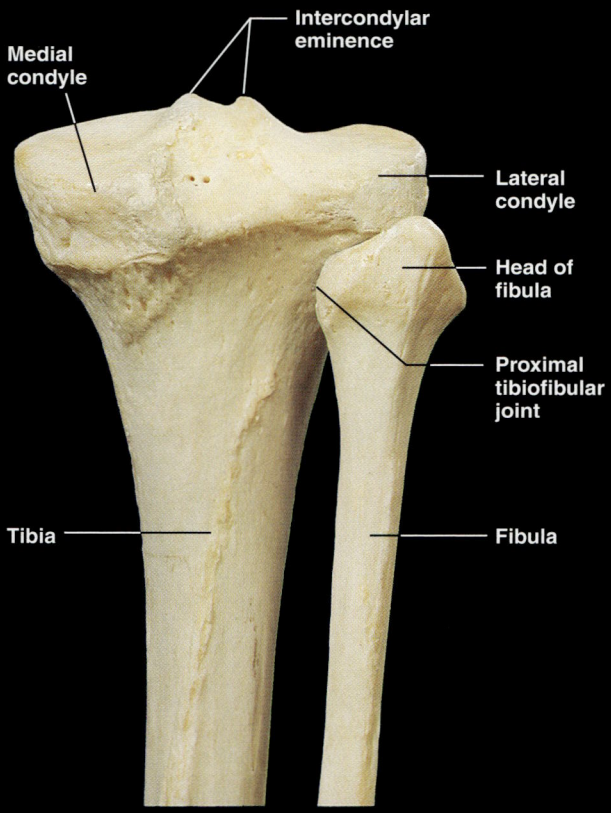

(f) articulated right tibia and fibula, proximal end, posterior view

Figure 30 Right tibia and fibula (continued

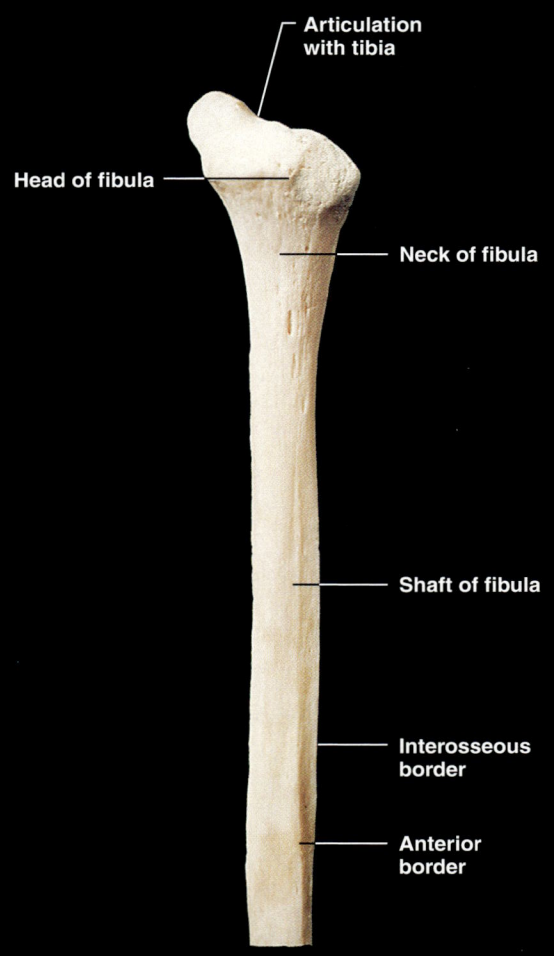

(h) right fibula, proximal end, anterior view

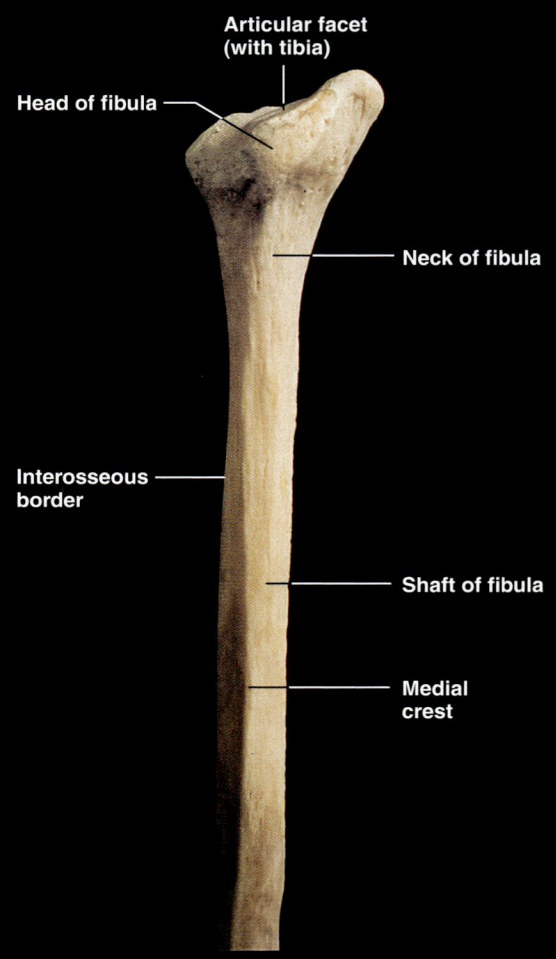

(i) right fibula, proximal end, posteromedial view

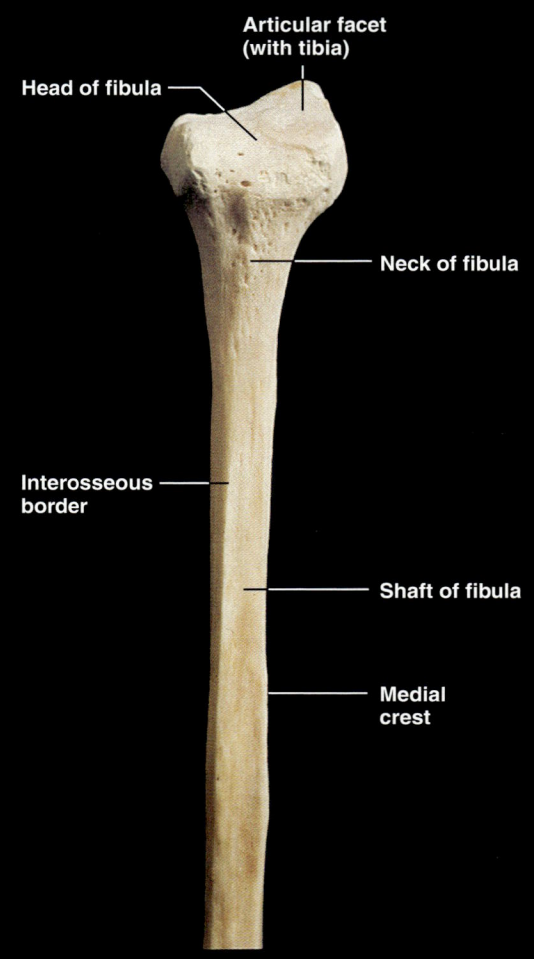

(j) right fibula, proximal end, medial view

Figure 30 Right tibia and fibula (continued).

Phalanges (toes)
Distal phalanx
Proximal phalanx
Metatarsal bones
1
2
3
4
5
Tarsals
Medial cuneiform
Intermediate cuneiform
Navicular
Talus
Calcaneus
Cuboid
Lateral cuneiform
Base
Body
Head
Proximal phalanx
Middle phalanx
Distal phalanx

(a) superior surface

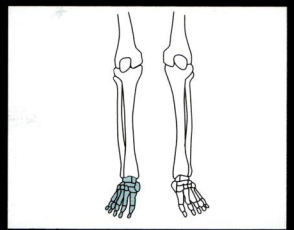

Distal phalanx
Middle phalanx
Proximal phalanx
Phalanges (toes)
Distal phalanx of hallux
Proximal phalanx of hallux
Metatarsal bones
Head and 2 sesamoid bones
Body
Base
5
4
3
2
1
Tarsals
Medial cuneiform
Intermediate cuneiform
Lateral cuneiform
Navicular
Cuboid
Sustentaculum tali (talar shelf)
Talus
Calcaneus
Calcaneal tuberosity

(b) inferior (plantar) surface

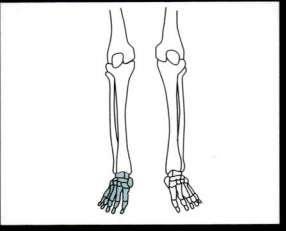

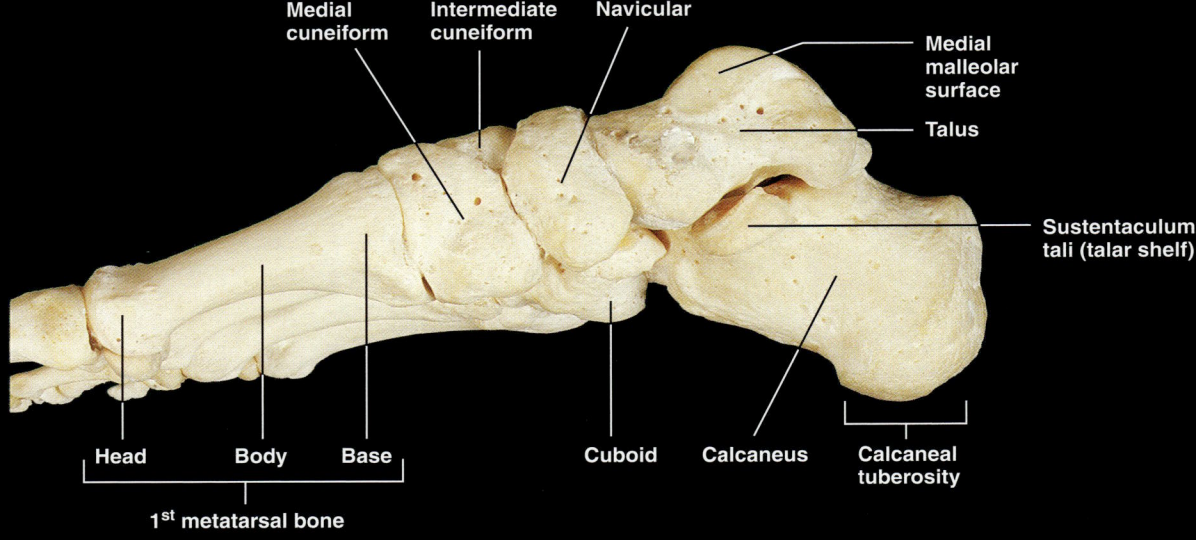

(c) medial view

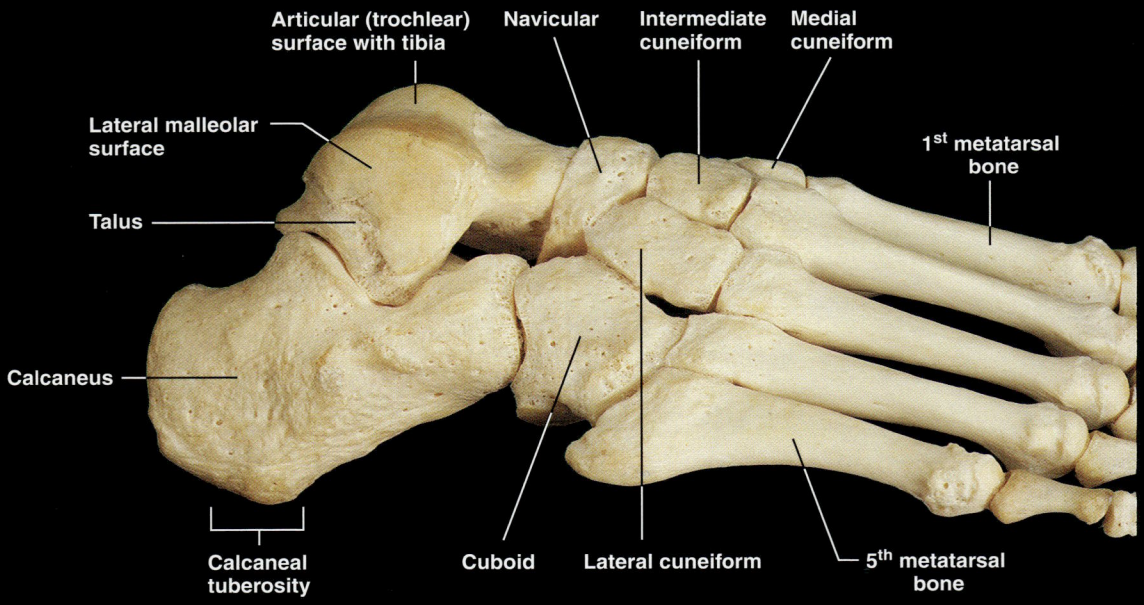

(d) lateral view

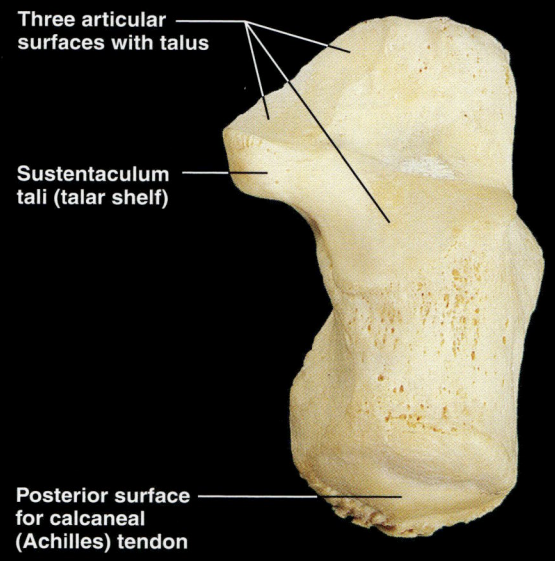

(e) right calcaneus, superior aspect

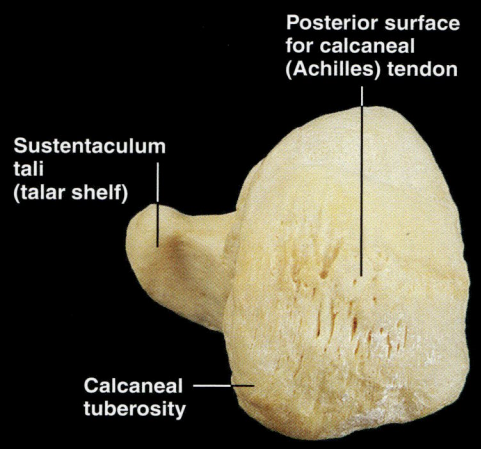

(f) right calcaneus, posterior aspect

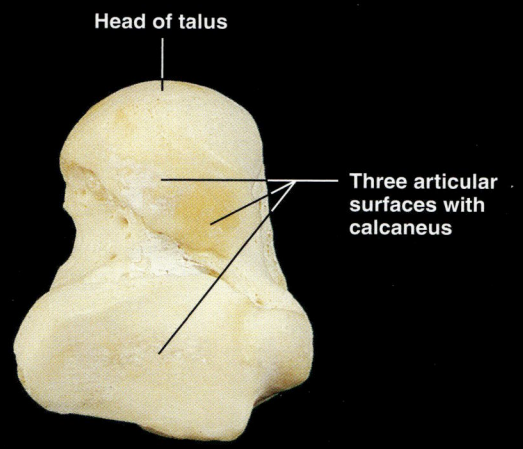

(g) right talus, inferior view

Figure 31 Bones of the right ankle and foot (continued).

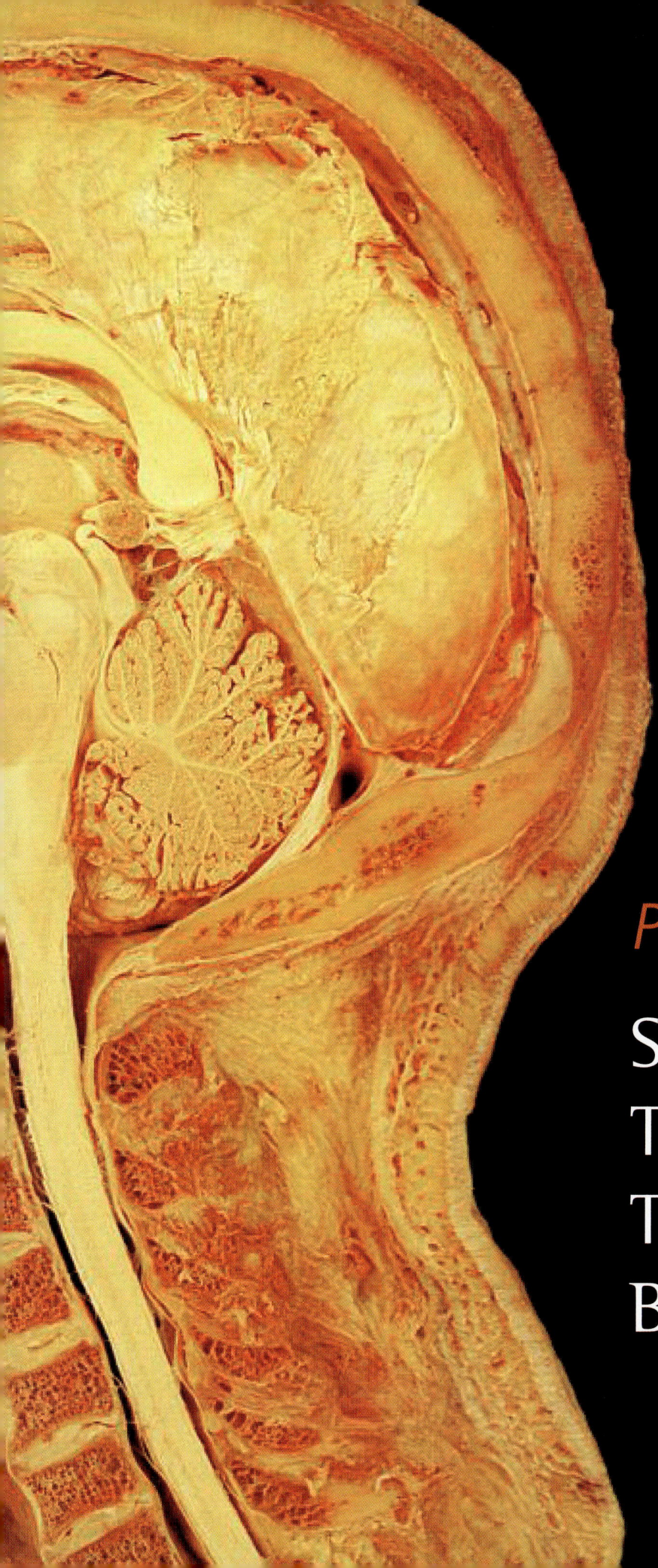

Part III

SOFT TISSUE OF THE HUMAN BODY

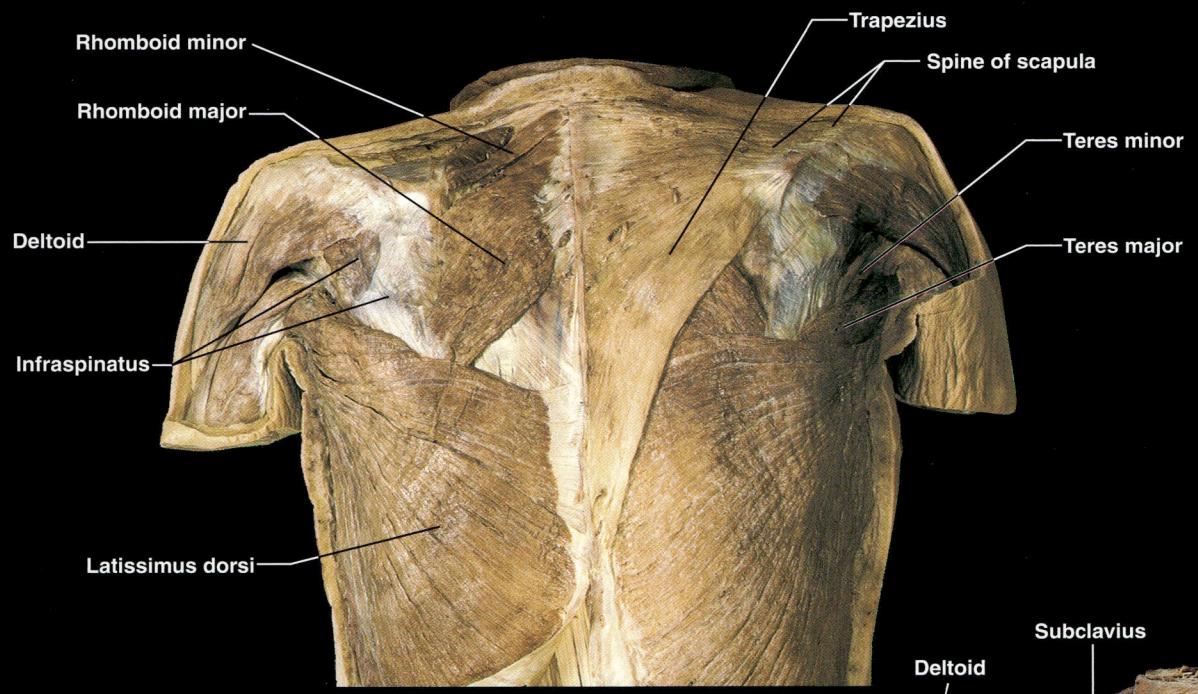

Figure 32 Superficial muscles of the thorax, posterior view.

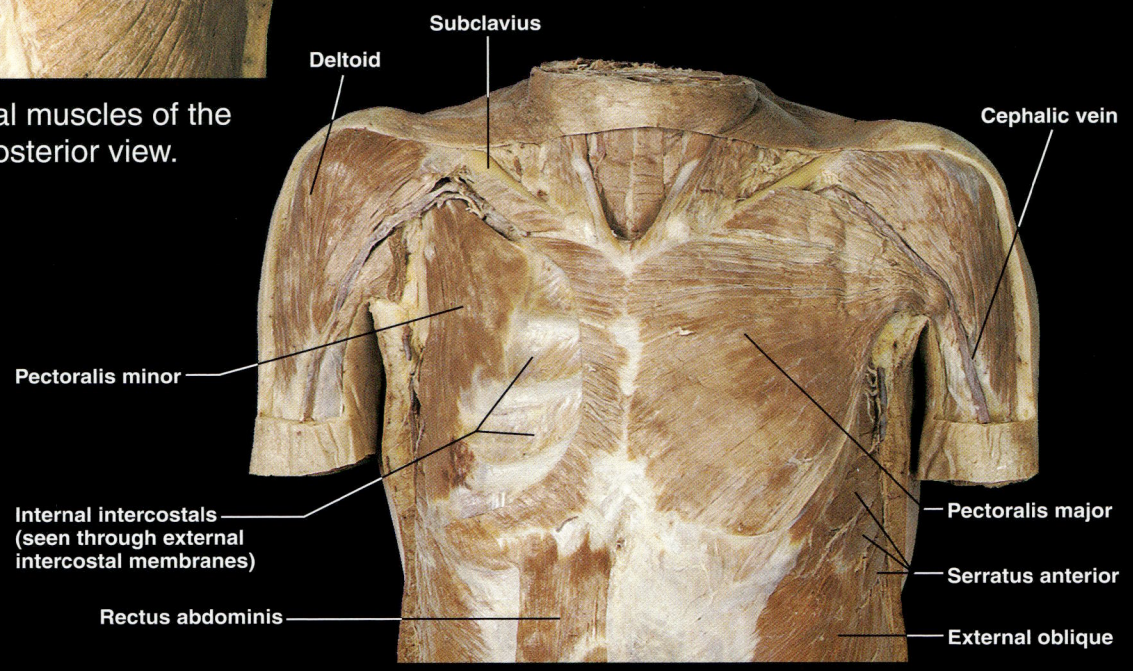

Figure 33 Superficial muscles of the thorax, anterior view.

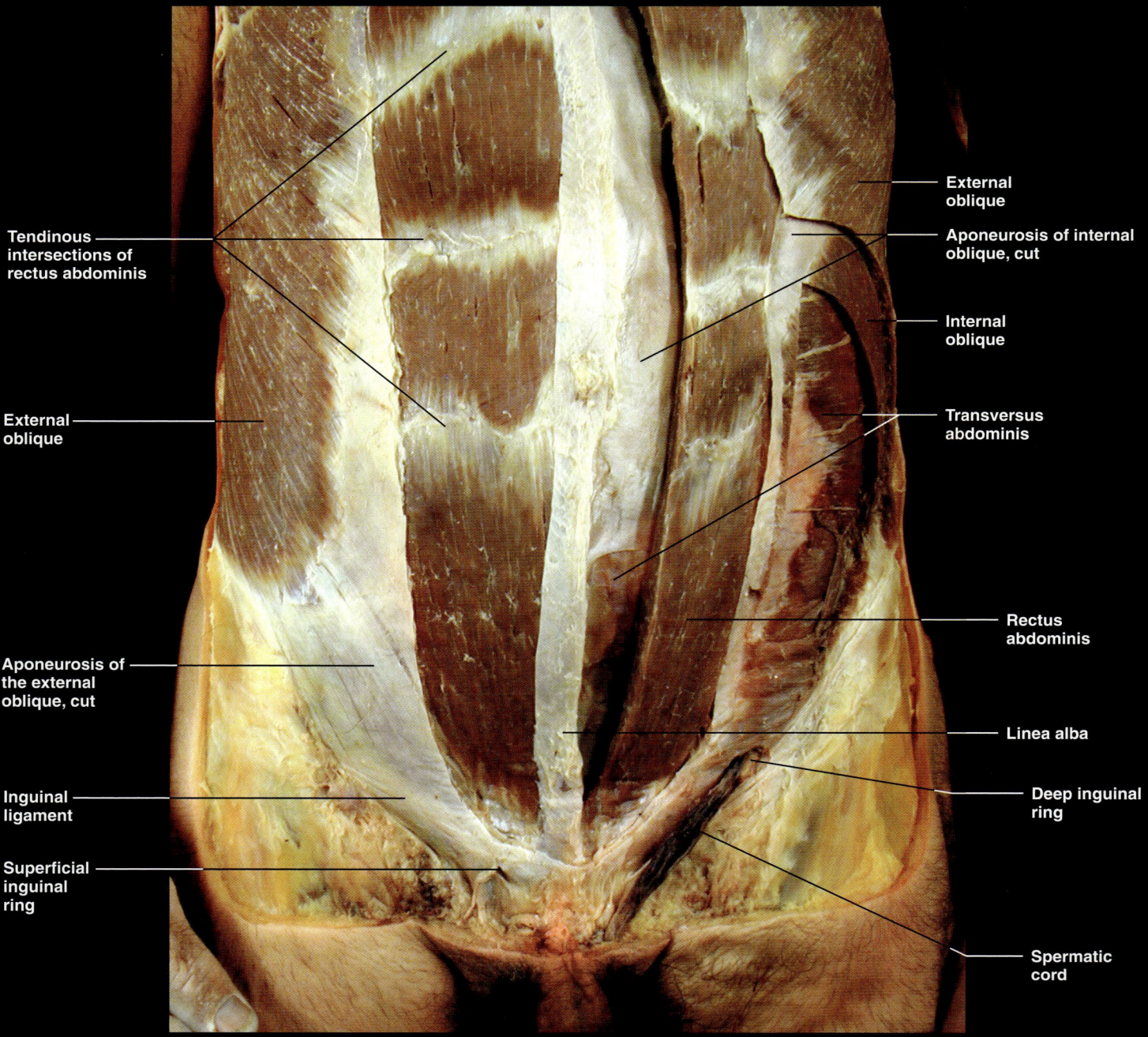

Figure 34 Abdominal muscles.

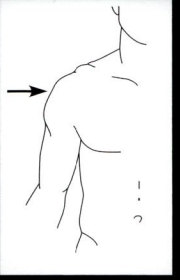

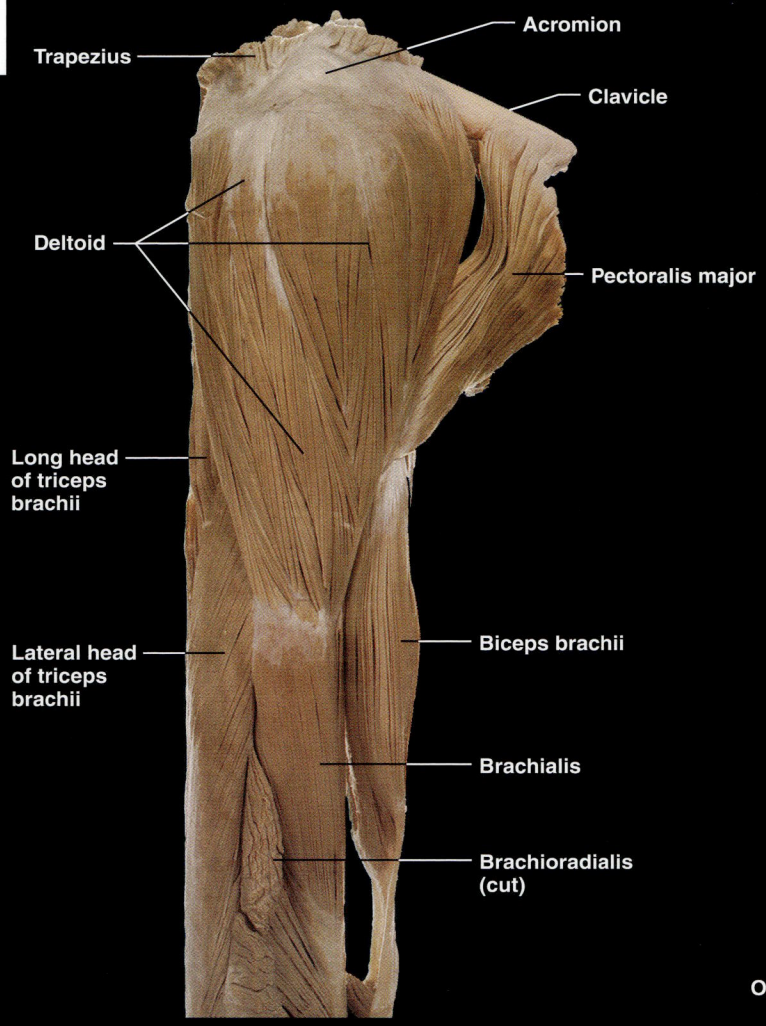

Figure 35 Right shoulder from right, showing deltoid muscle and biceps.

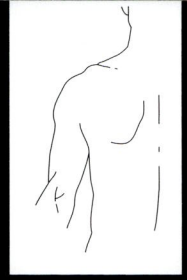

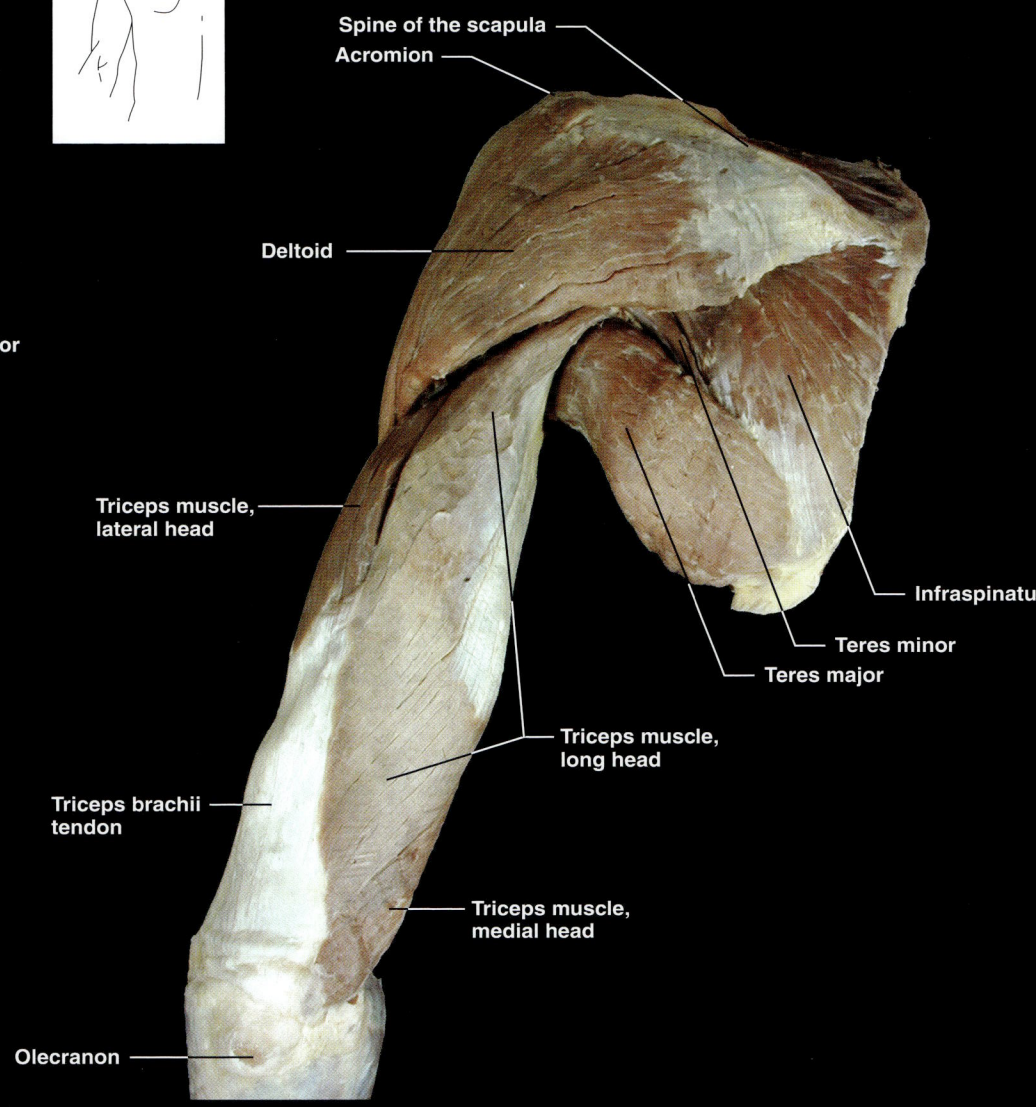

Figure 36 Triceps of the left arm, posterior view.

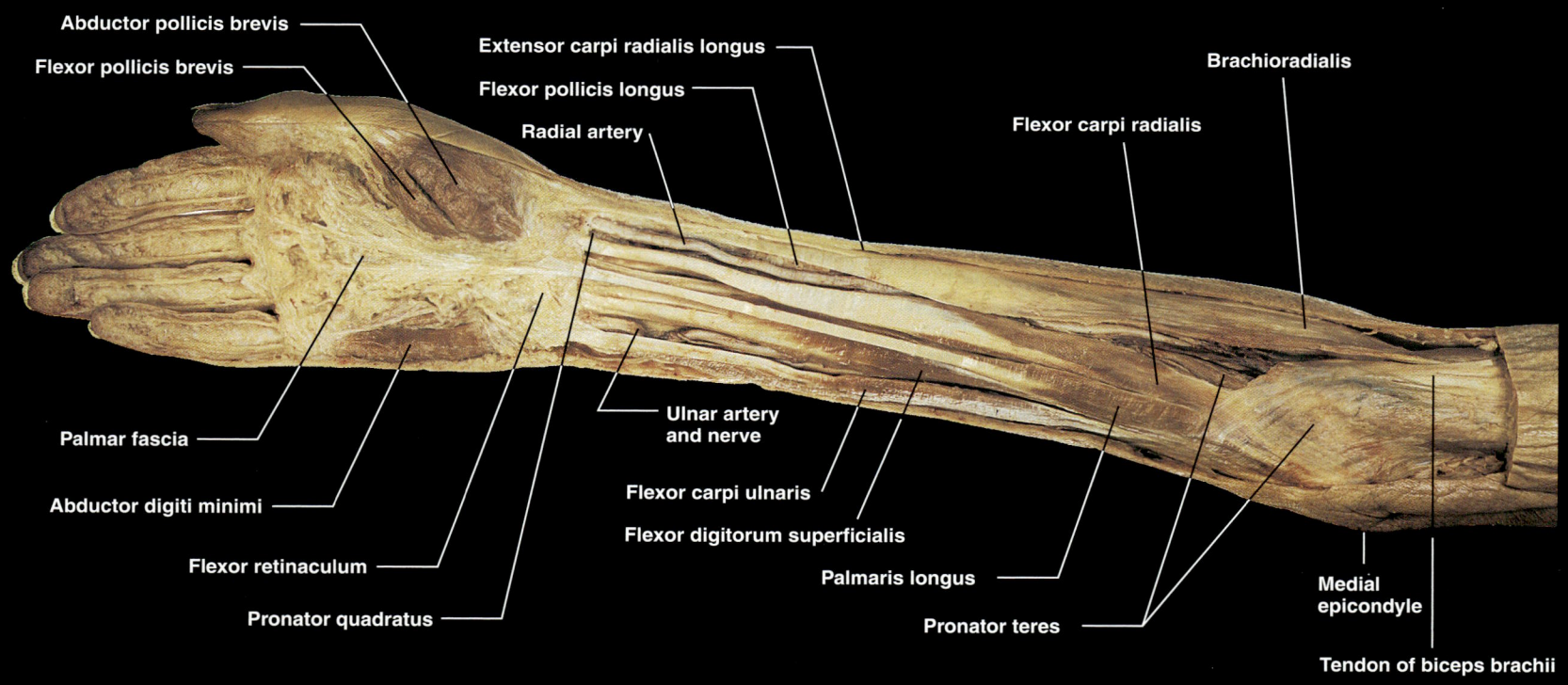

(a) palmar surface

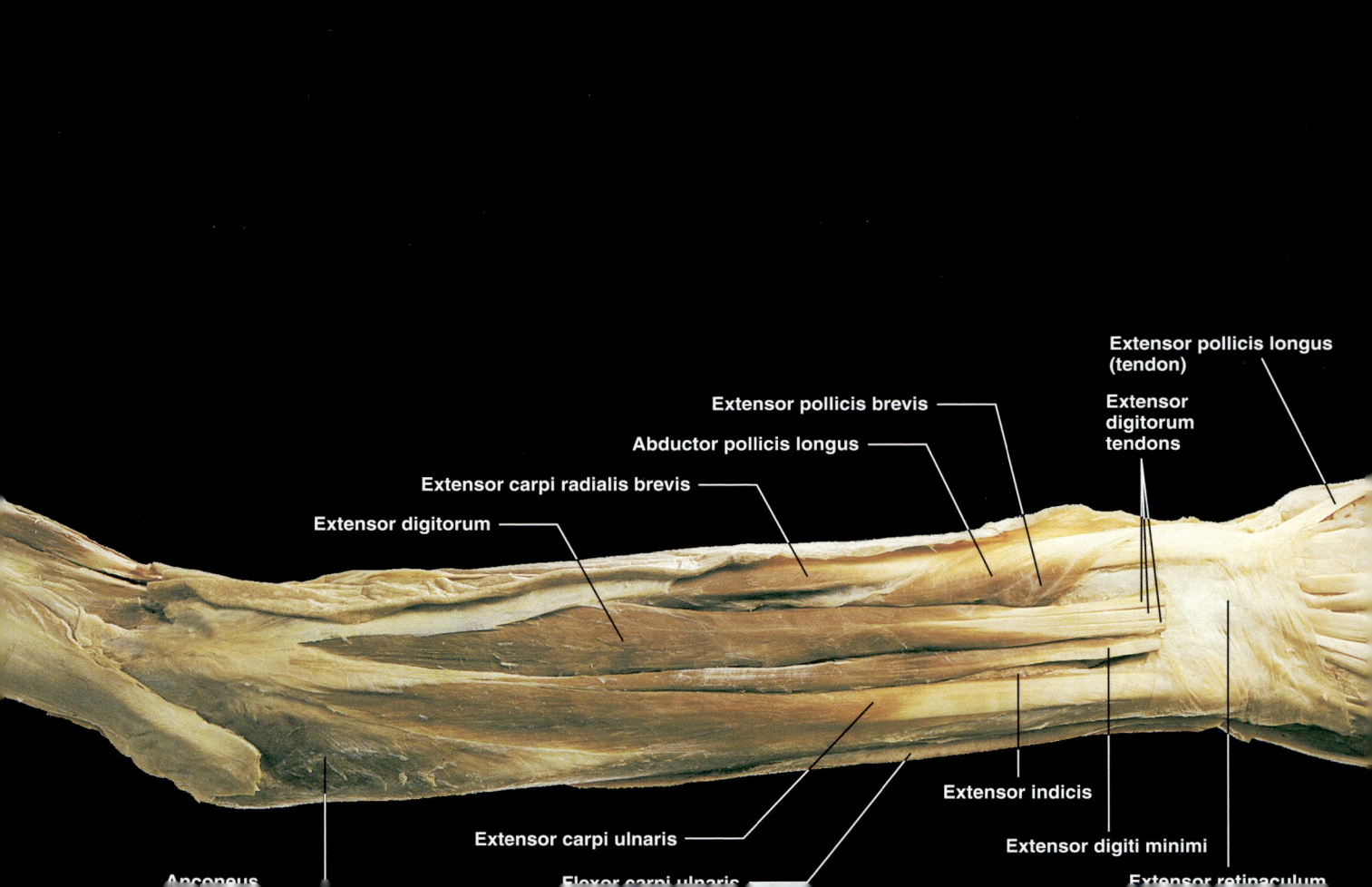
Extensor pollicis longus (tendon)
Extensor pollicis brevis
Extensor digitorum tendons
Abductor pollicis longus
Extensor carpi radialis brevis
Extensor digitorum
Extensor indicis
Extensor carpi ulnaris
Extensor digiti minimi
Anconeus
Flexor carpi ulnaris
Extensor retinaculum

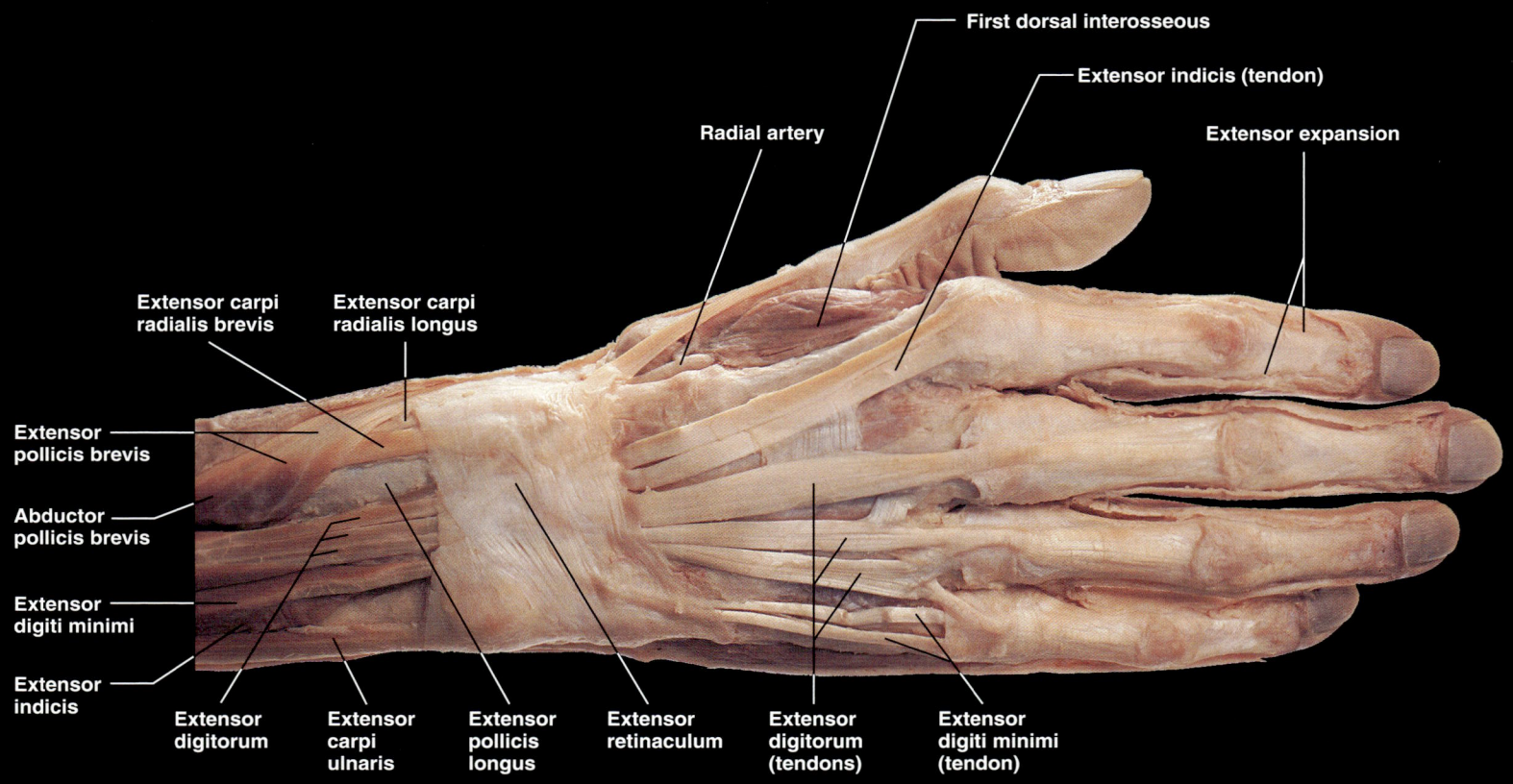

(a) dorsum surface of the right hand and wrist

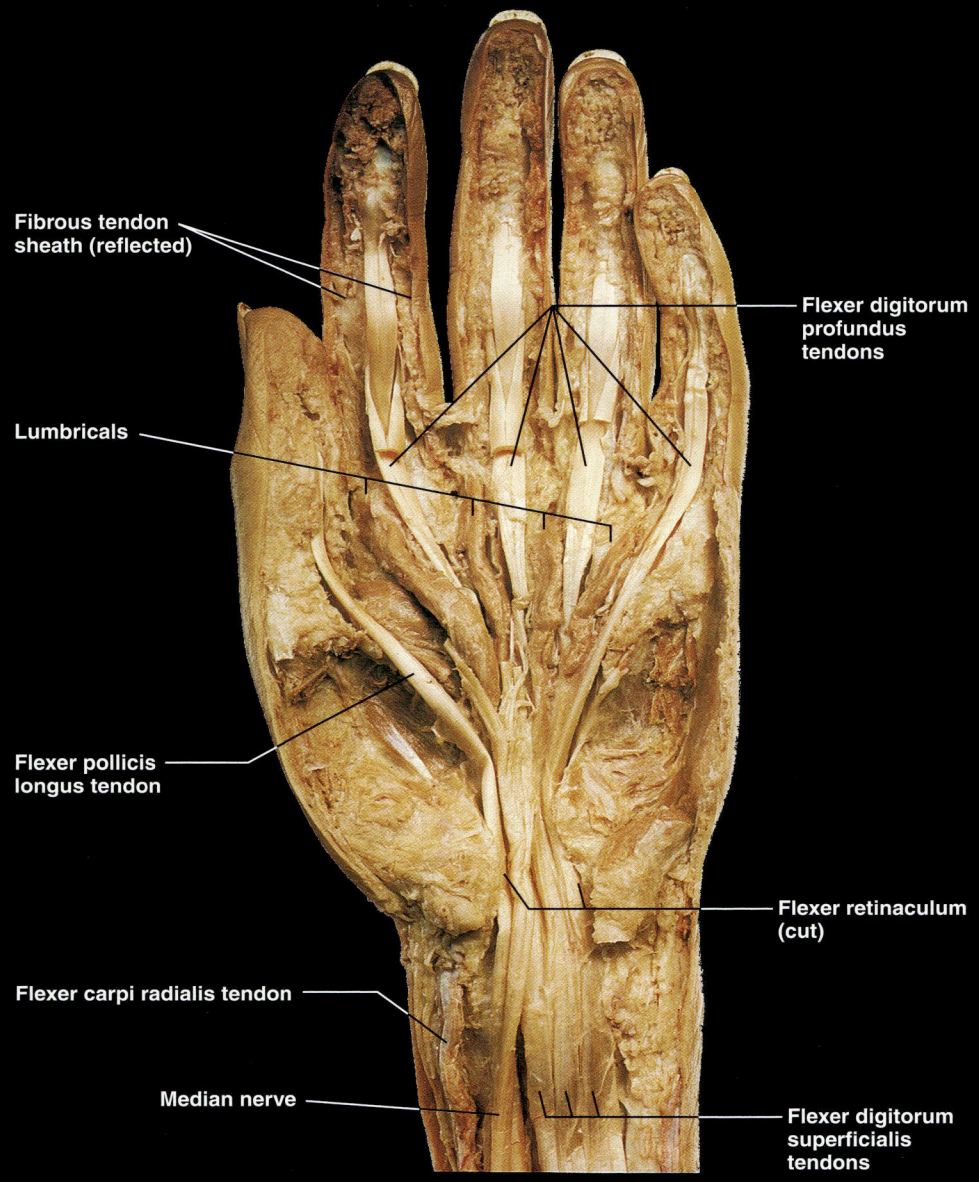

(b) palmar surface of the left hand and wrist

Figure 38 Wrist and hand.

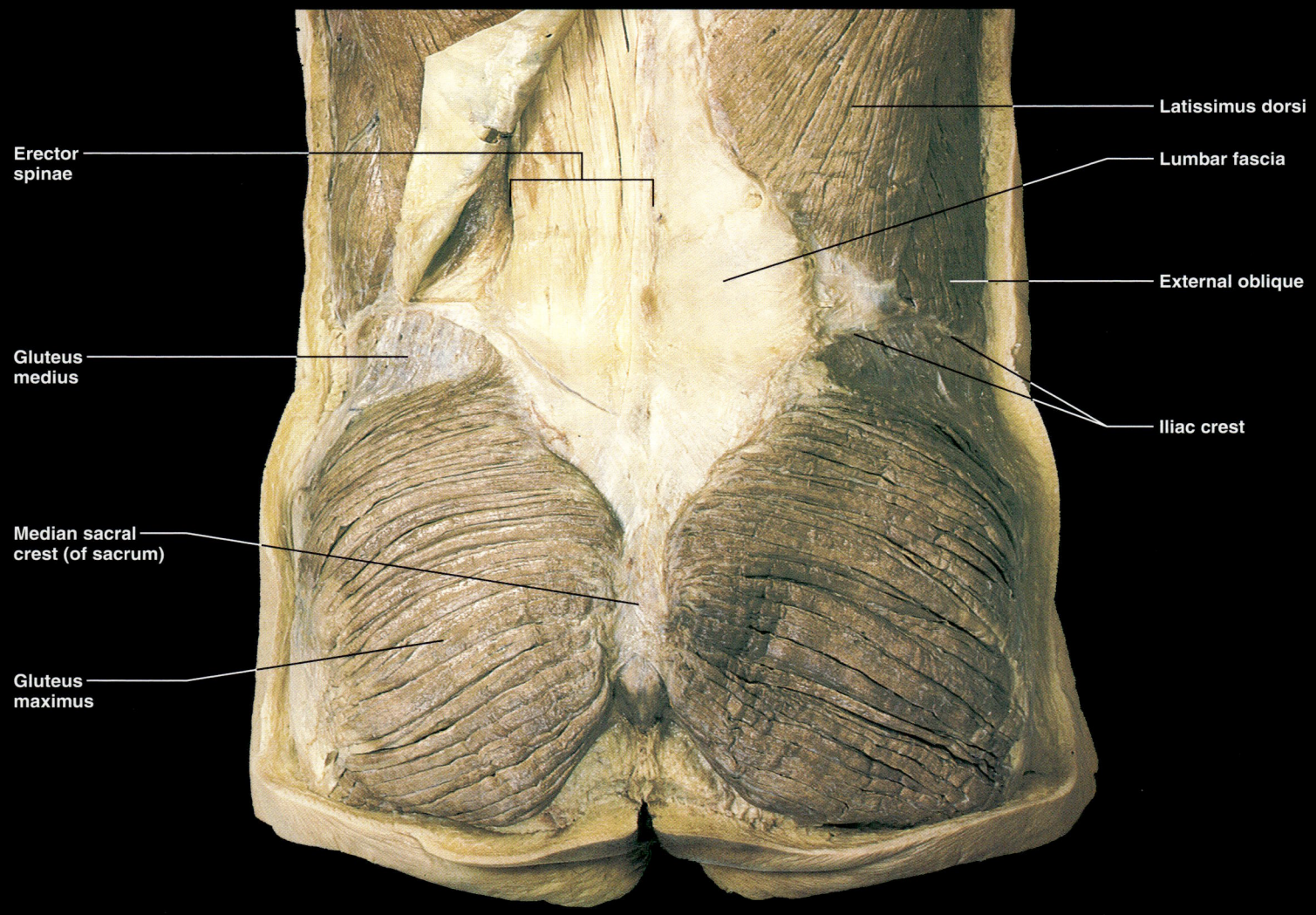

Figure 39 Superficial muscles of the superior gluteal region.

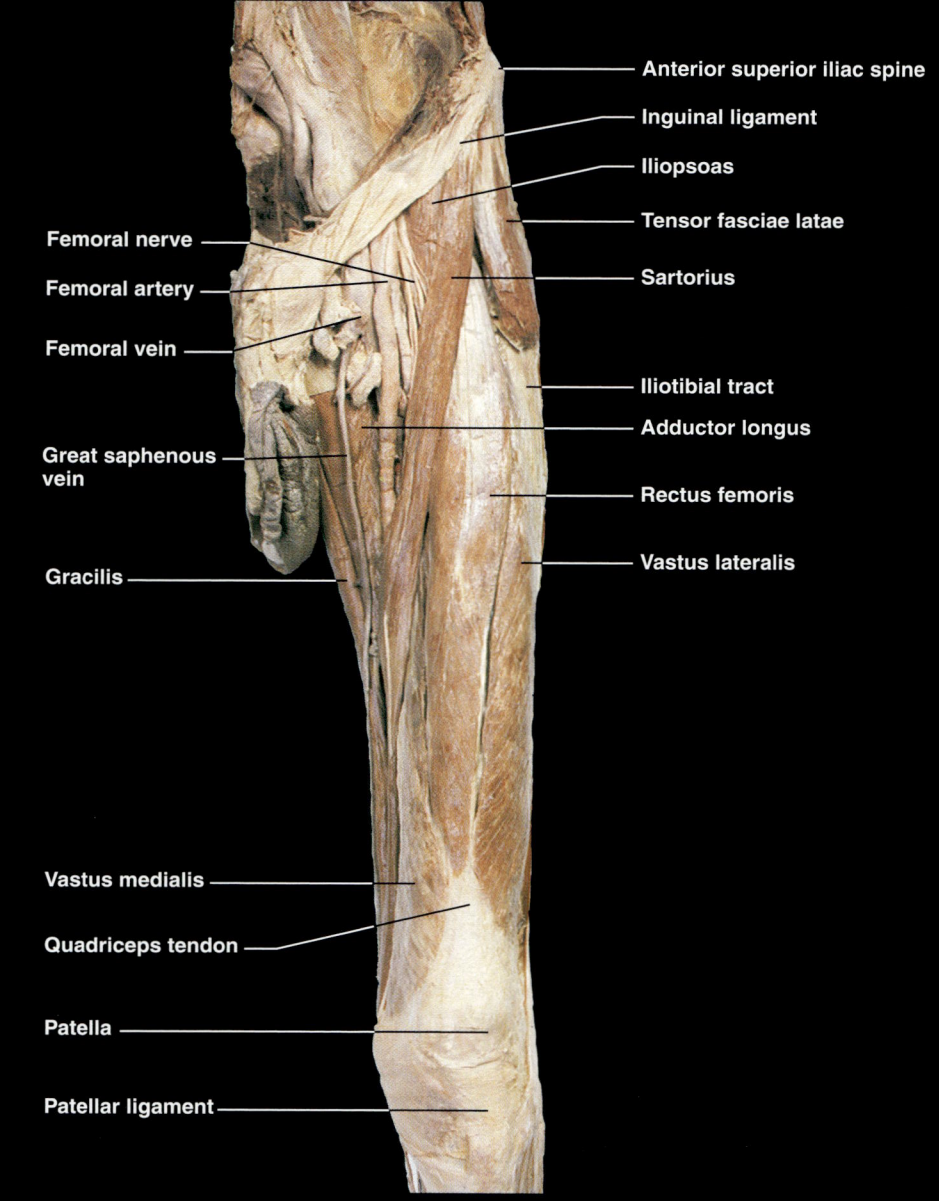

Figure 40 Superficial muscles of the left lower thigh, anterior view.

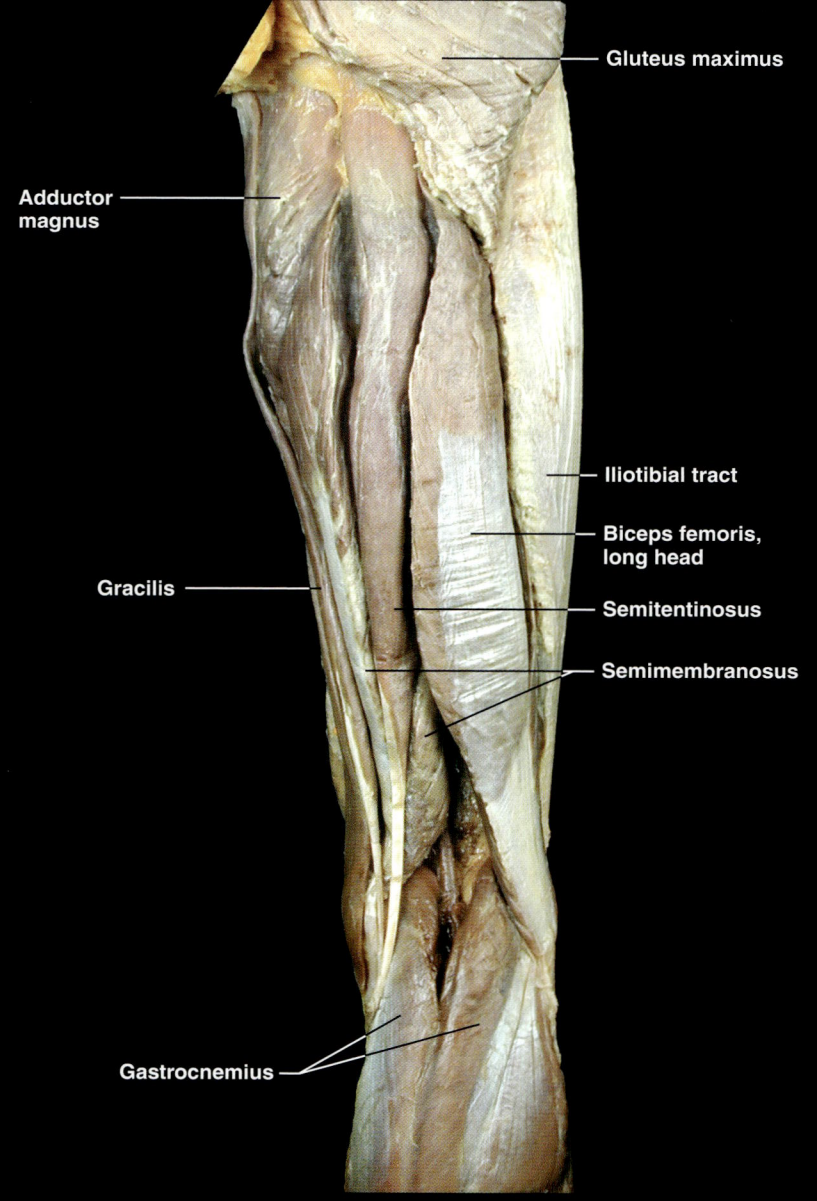

Figure 41 Right upper thigh, posterior view.

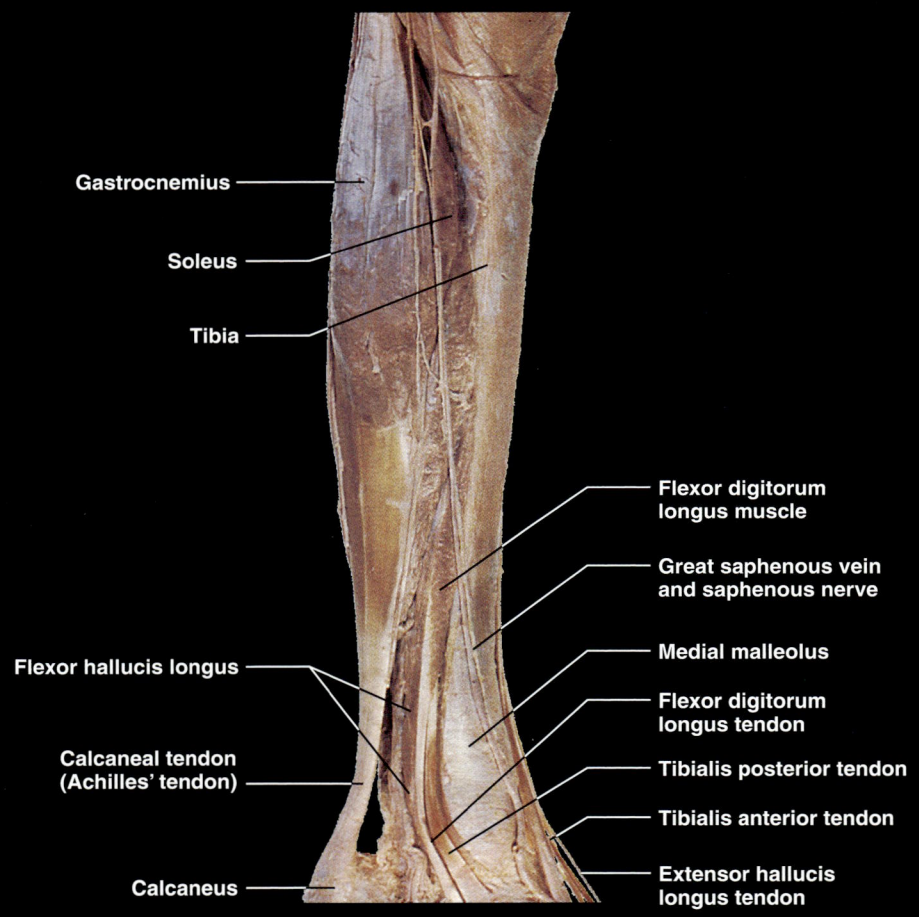

(a) medial view

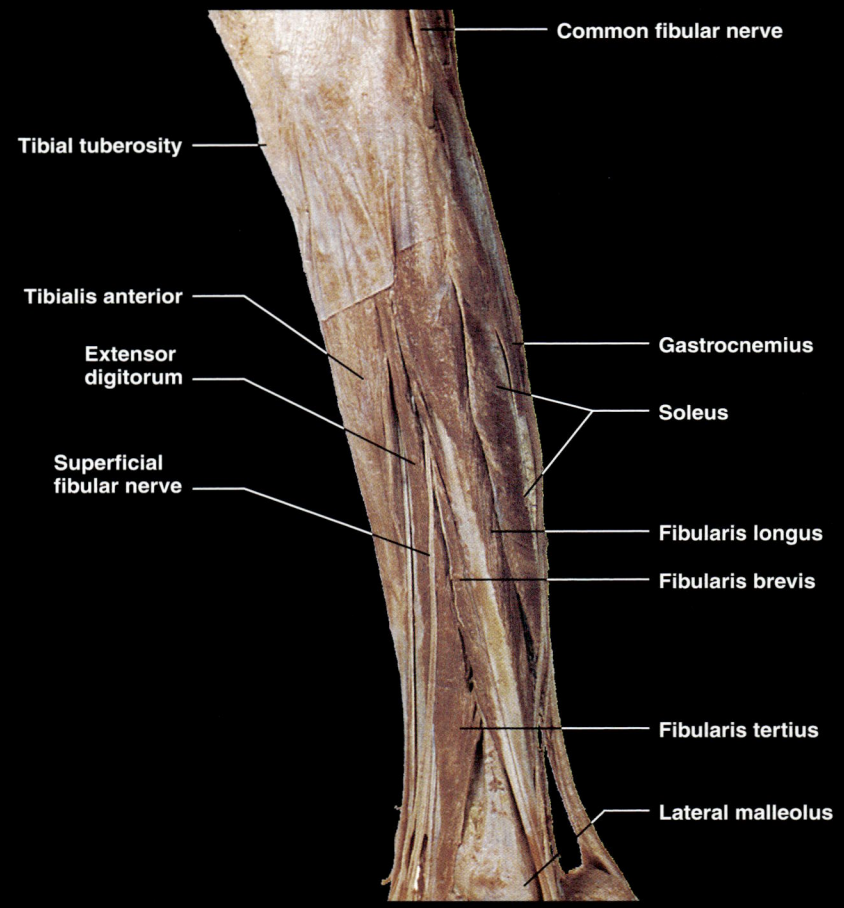

(b) lateral view

Figure 42 Leg.

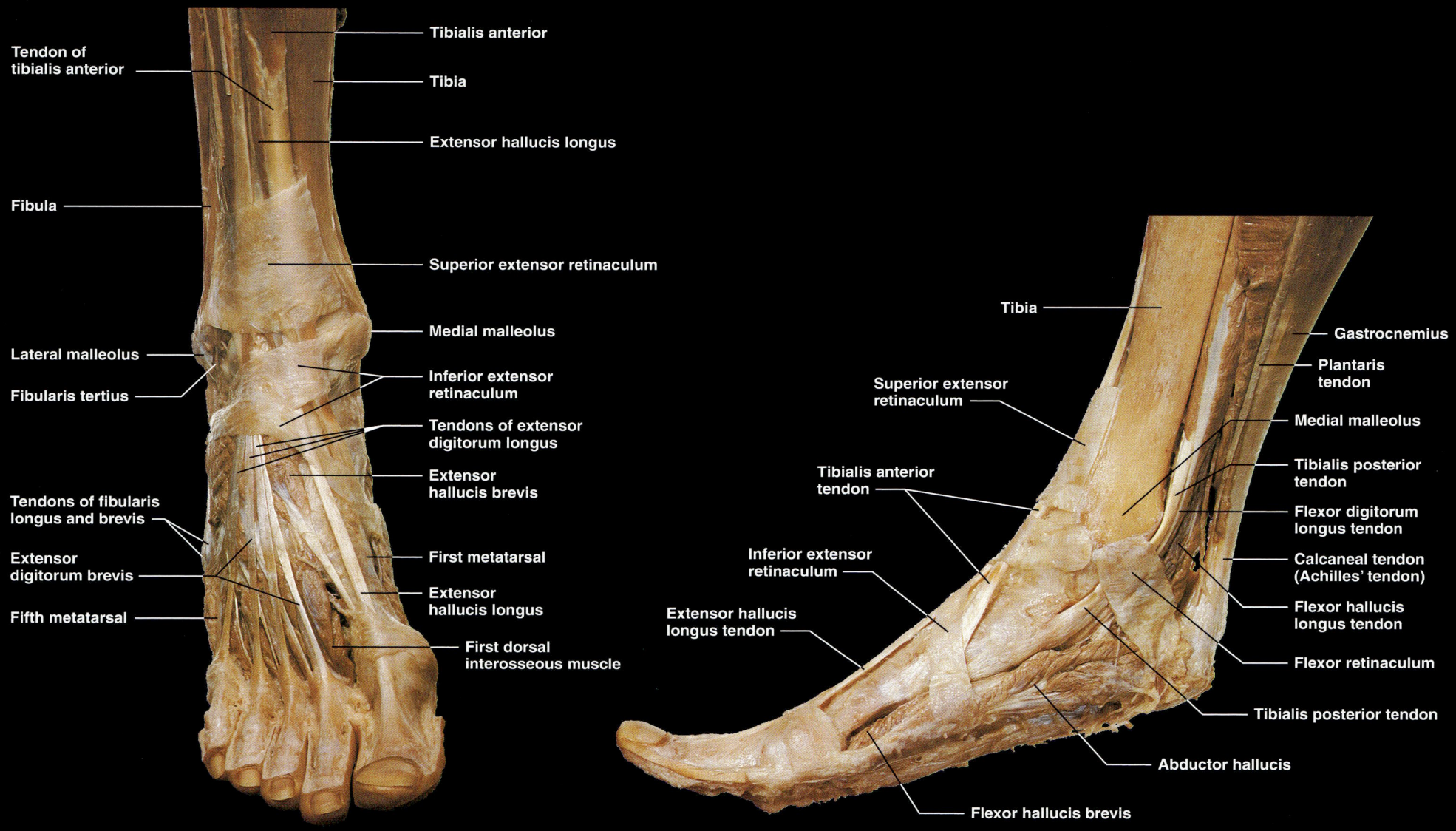

(a) anterior view

(b) medial view

Figure 43 Foot.

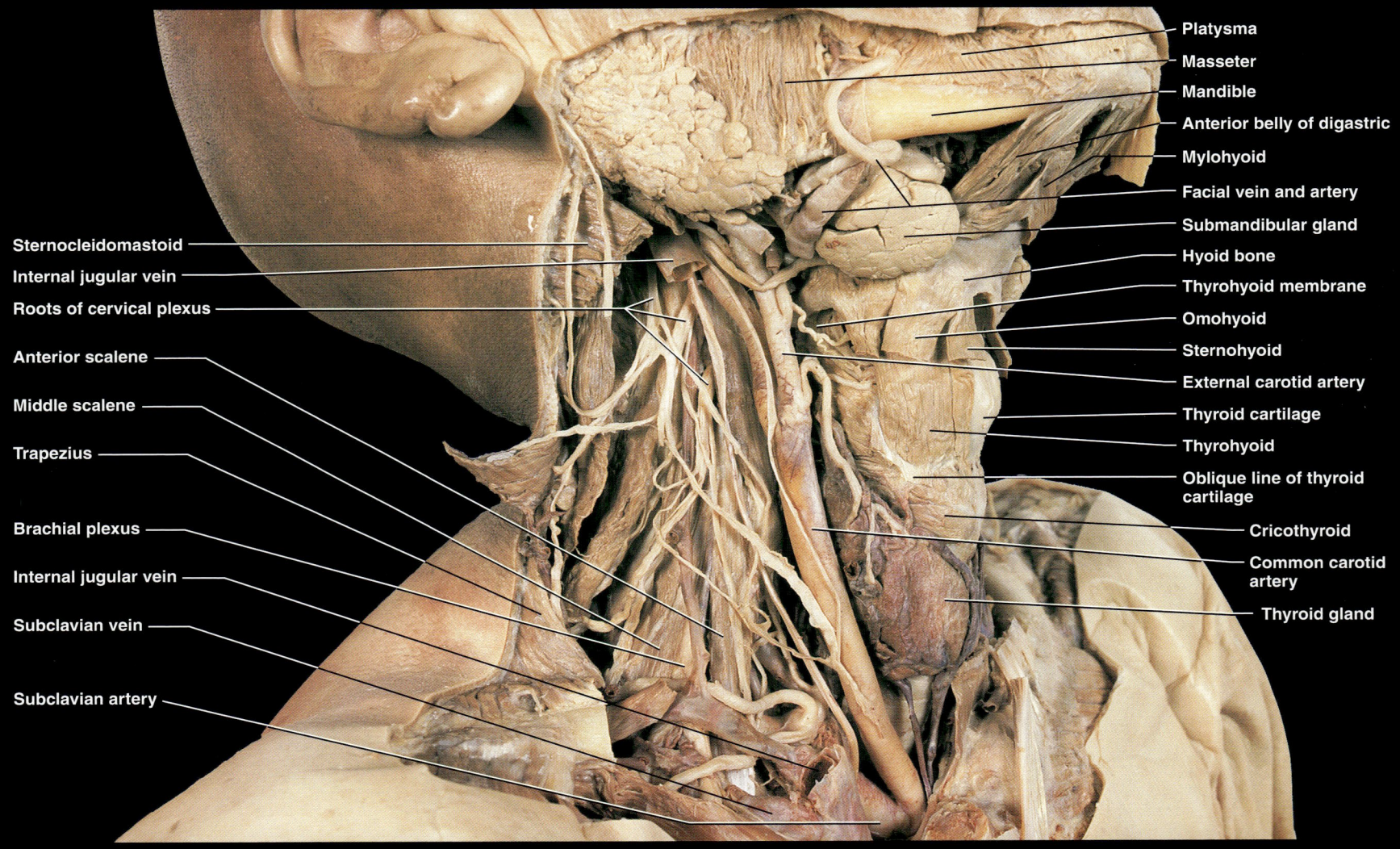

Figure 44 Right lower face and upper neck.

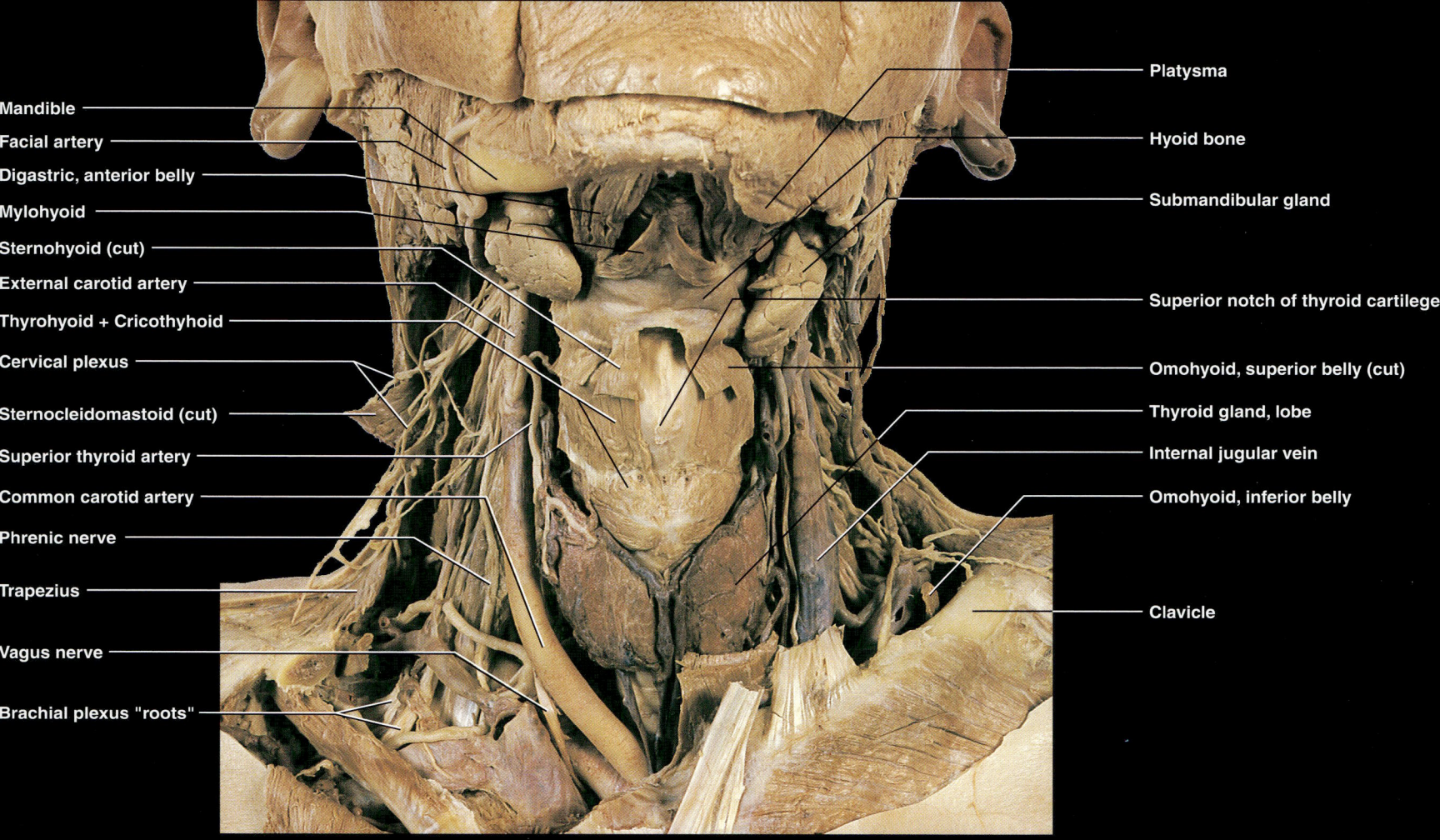

Figure 45 Muscles, blood vessels, and nerves of neck, anterior view.

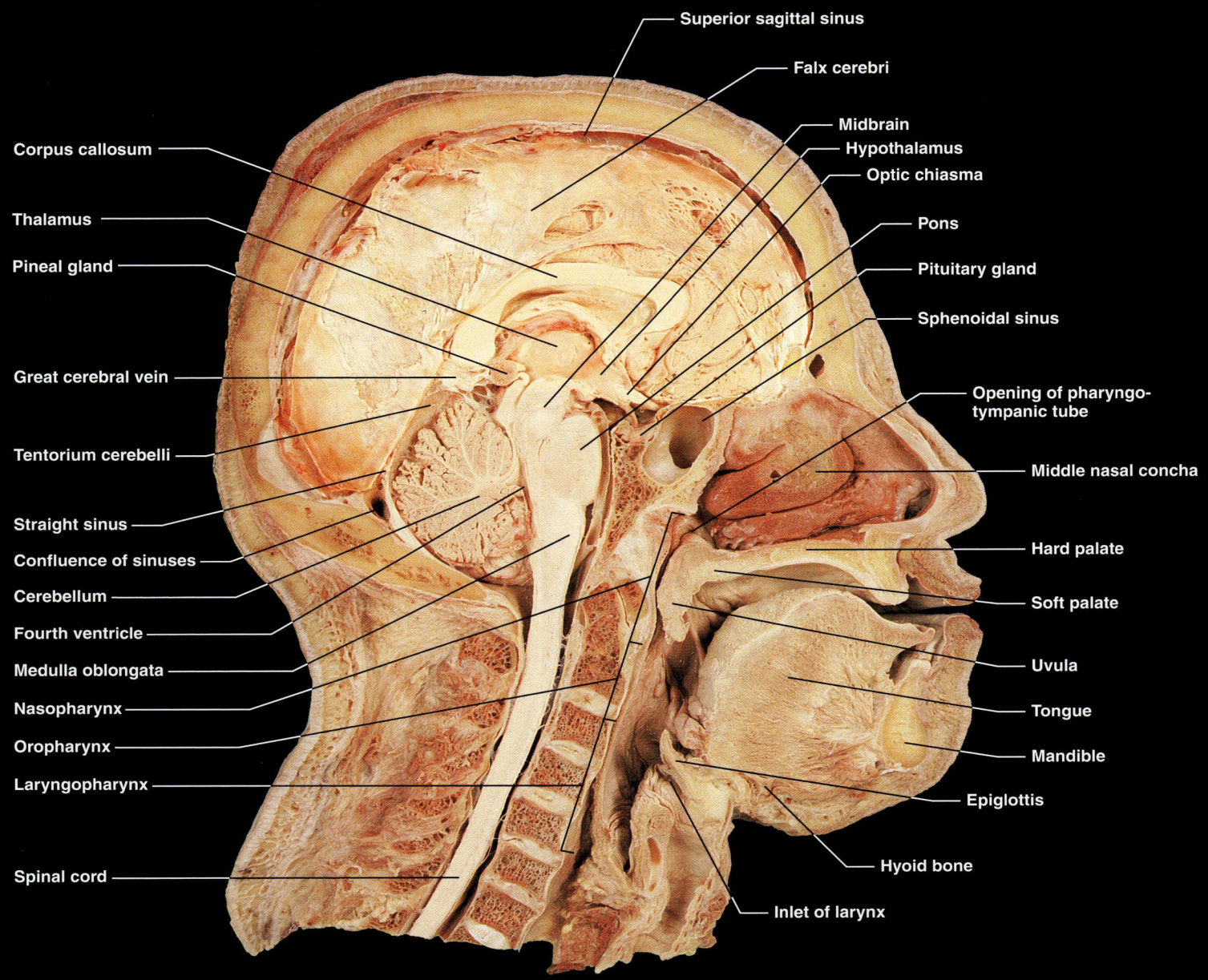

Figure 46 Sagittal section of the head.

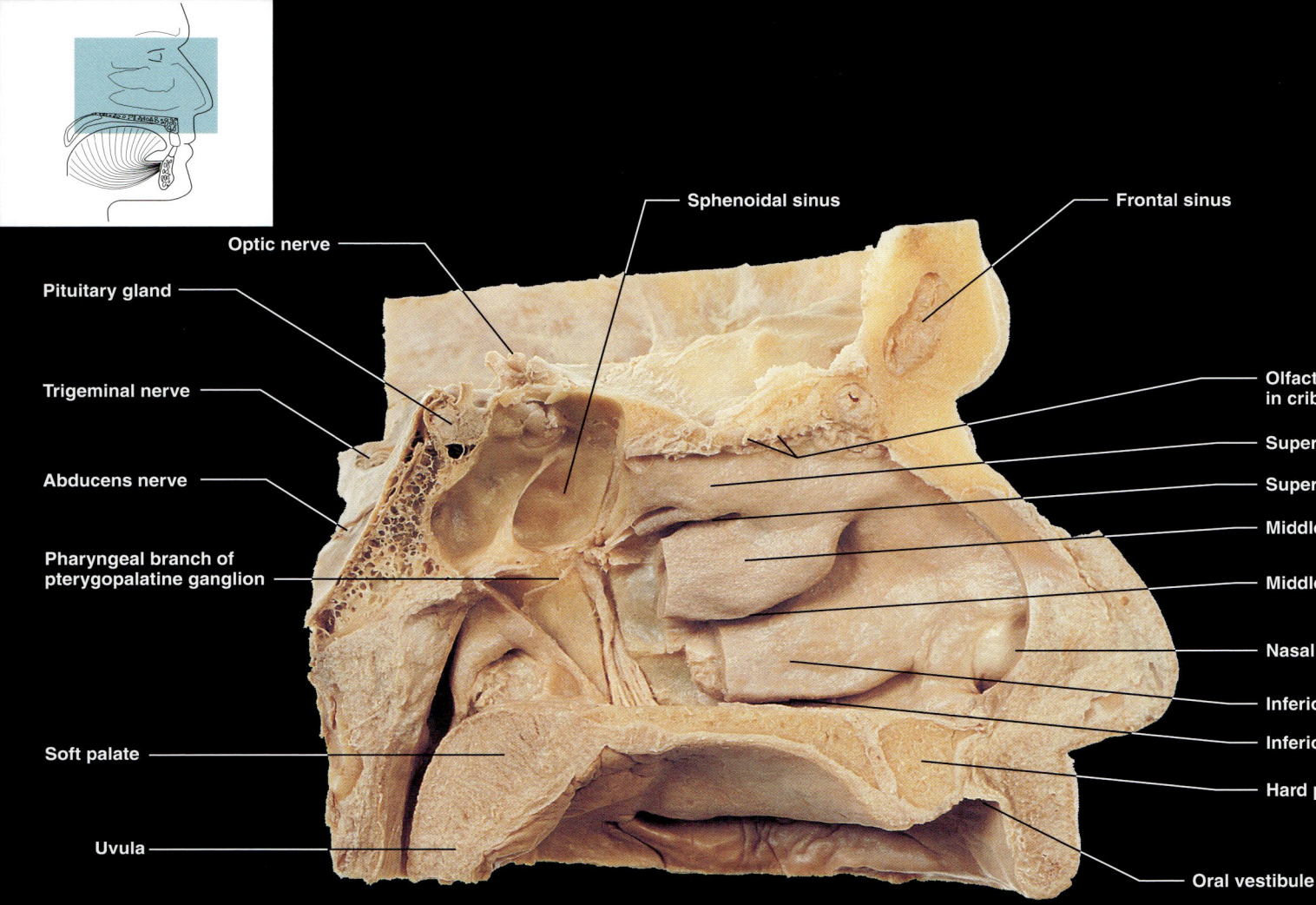

Figure 47 Left nasal cavity, lateral wall.

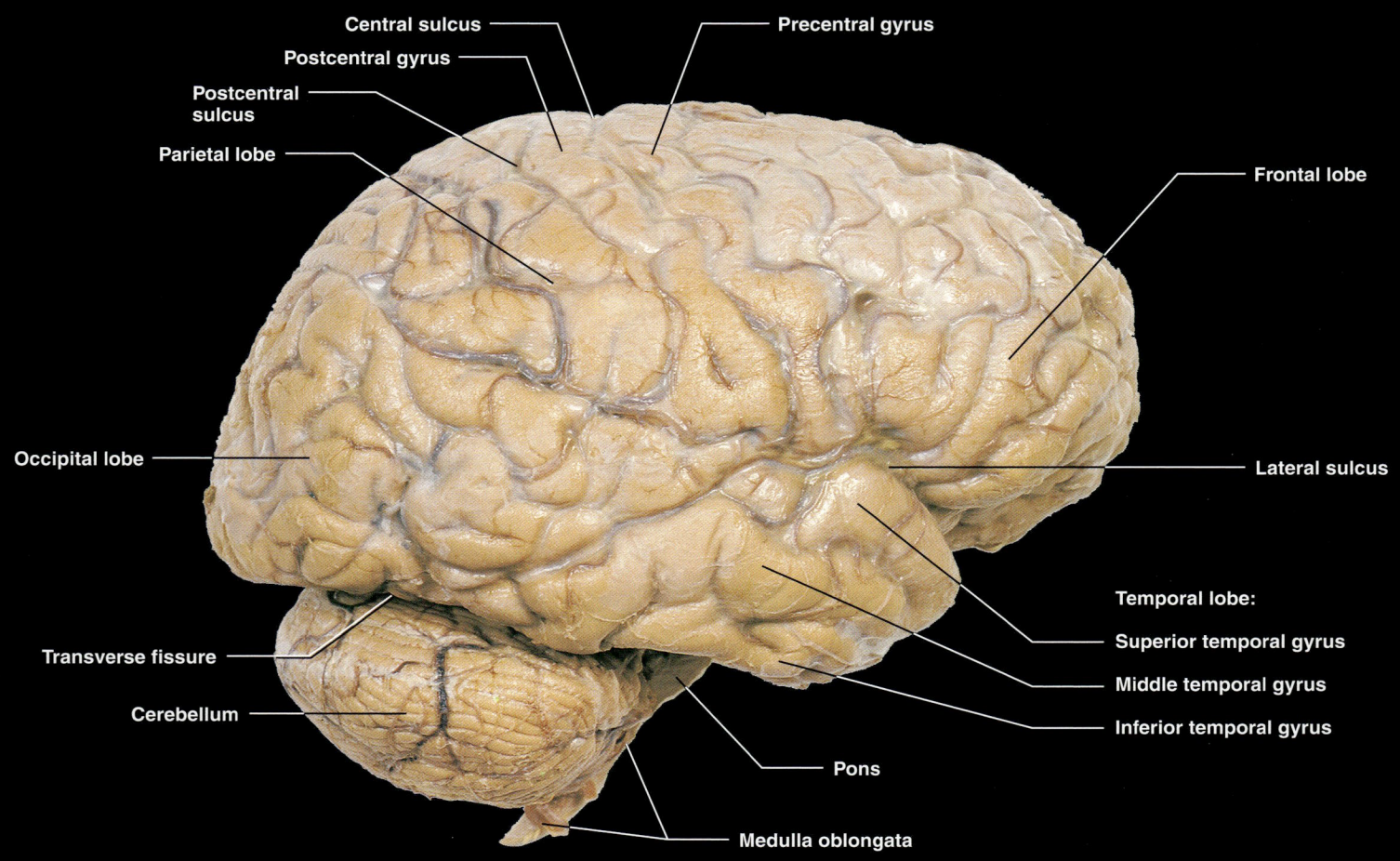

Figure 48 Right cerebral hemisphere (arachnoid mater removed).

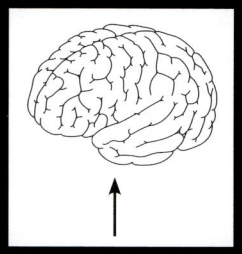

Frontal lobe

Optic nerve (II)

Optic chiasma

Infundibulum of pituitary gland

Mammillary body

Basilar artery

Pons

Choroid plexus of fourth ventricle

Medulla oblongata

Olfactory bulb

Olfactory tract

Internal carotid artery

Optic tract

Oculomotor nerve (III)

Trochlear nerve (IV)

Trigeminal nerve (V)

Abducens nerve (VI)

Facial nerve (VII)

Vestibulocochlear nerve (VIII)

Glossopharyngeal nerve (IX)

Vagus nerve (X)

Accessory nerve (XI)

Cerebellum

Hypoglossal nerve roots (XII)

Olive of medulla oblongata

Pyramid of medulla oblongata

Figure 49 Ventral view of the brain.

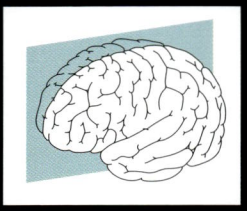

Cingulate gyrus
Fornix
Central sulcus
Hypothalamus sulcus
Lateral ventricle
Parieto-occipital sulcus
Corpus callosum
Third ventricle
Anterior commissure
Calcarine sulcus
Superior colliculus
Inferior colliculus
Corpora quadrigemina
Hypothalamus
Arbor vitae
Optic nerve and chiasma
Cerebellum
Mammillary body
Fourth ventricle
Uncus
Thalamus
Medulla oblongata
Pons

Figure 50 Midsagittal section of the brain.

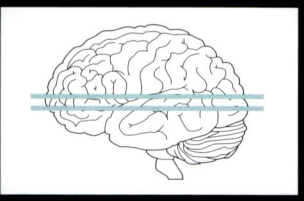

ANTERIOR

Anterior horn of lateral ventricle

Cerebral gray matter

Cerebral white matter

Head of caudate nucleus

Interventricular foramen

Corpus callosum

Internal capsule

Lentiform nuclei

Body of caudate nucleus

Insula

Body of fornix

Thalamus

Inferior horn of lateral ventricle

Choroid plexus

Posterior horn of lateral ventricle

Corpus callosum

Visual area of cerebral cortex

POSTERIOR

Figure 51 Transverse section of the brain, superior view. Left: on a level with the intraventricular foramen; right: about 1.5 cm higher.

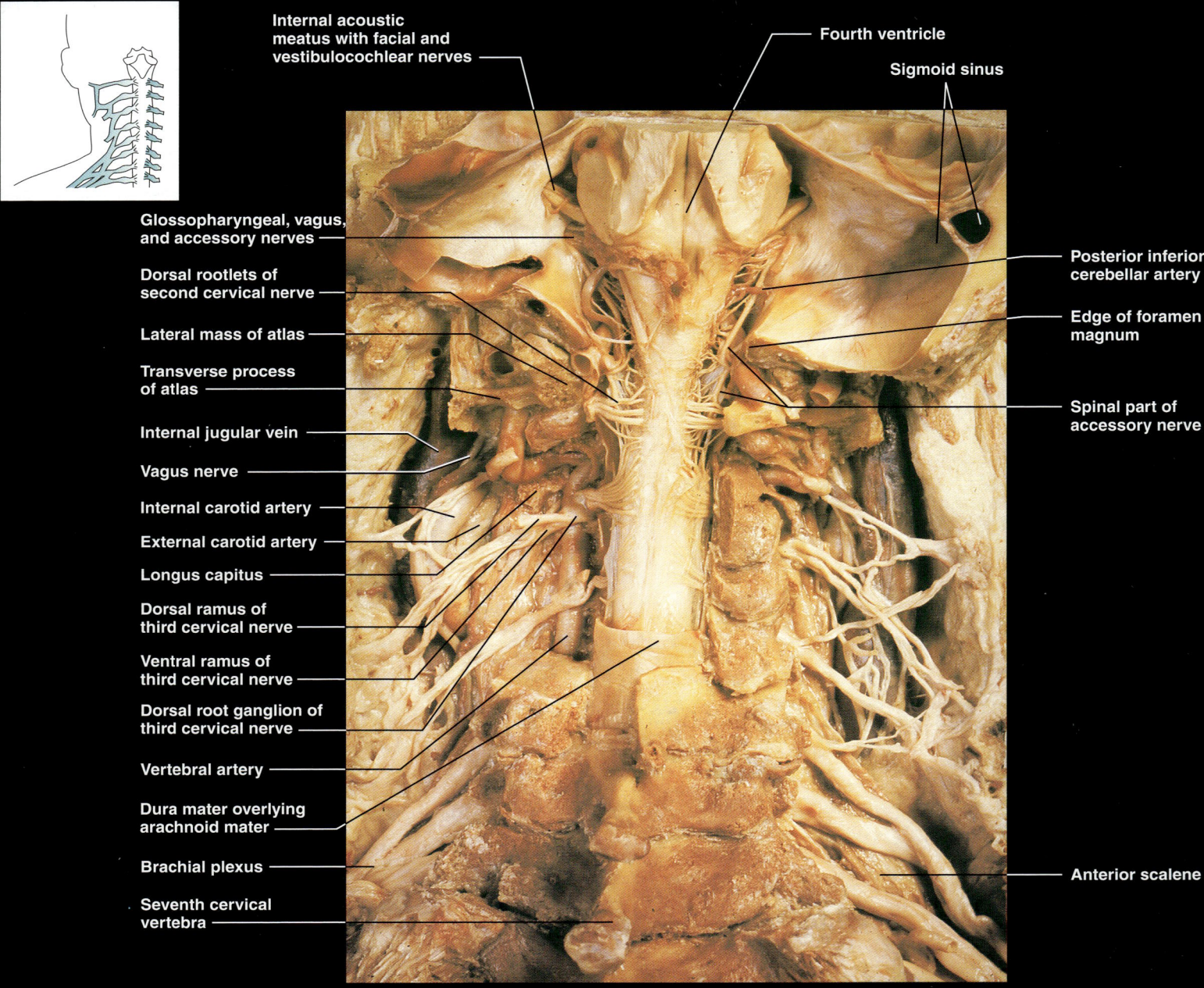

Figure 52 Brainstem and cervical region of the spinal cord, posterior view.

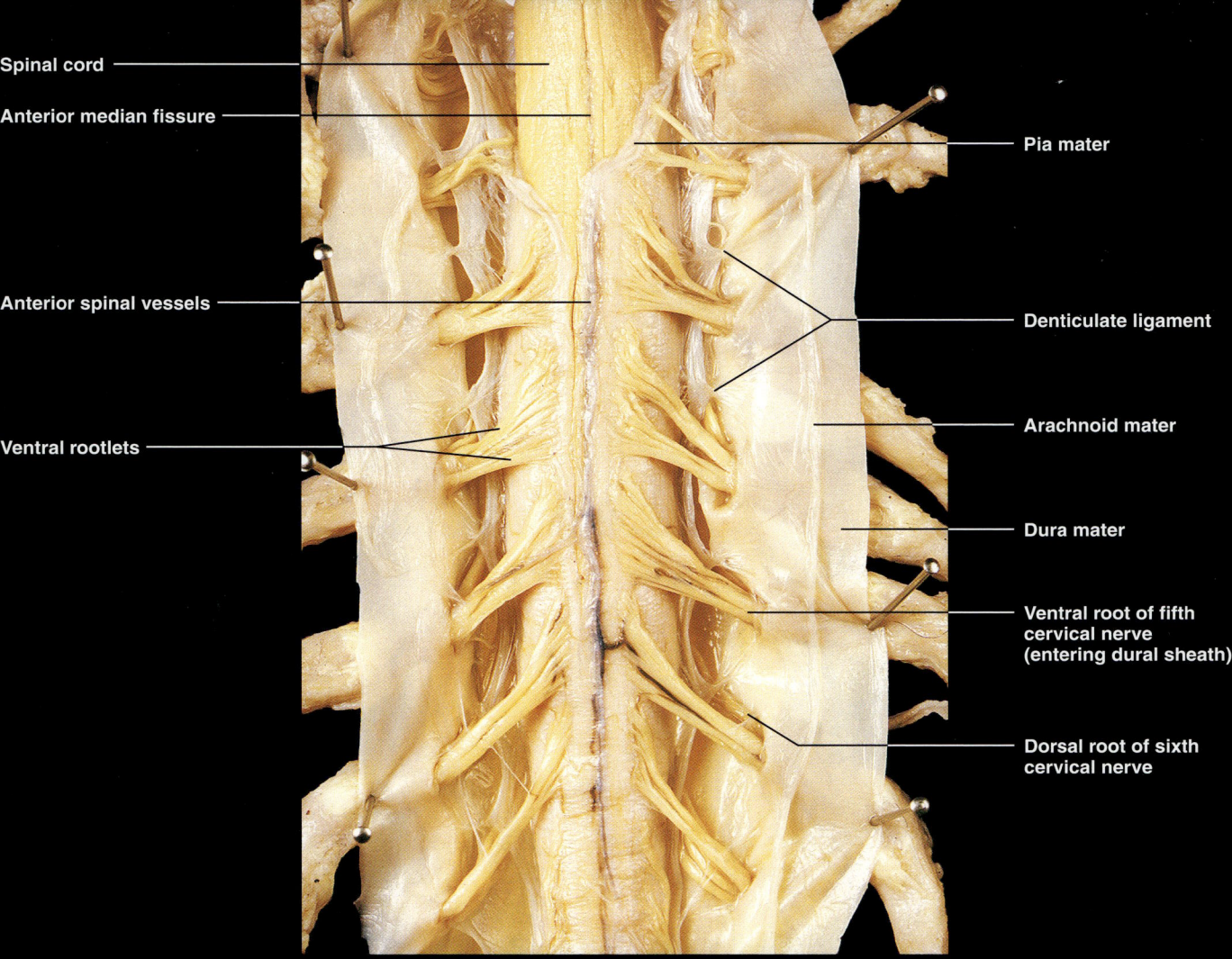

Figure 53 Cervical region of spinal cord, ventral view.

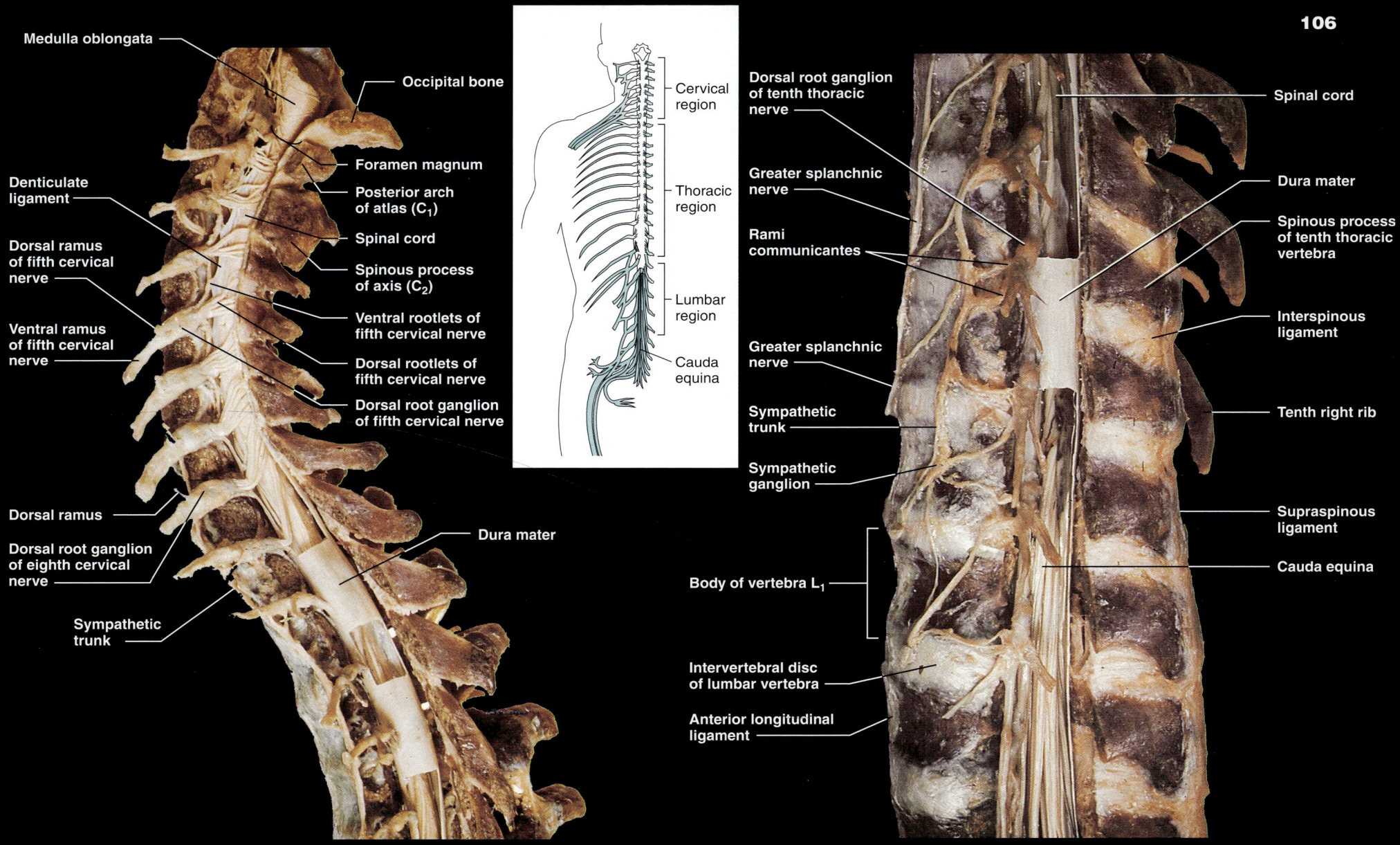

(a) cervical and upper thoracic regions from the left

(b) lower thoracic and upper lumbar regions from the left

Figure 54 Vertebral column and spinal cord.

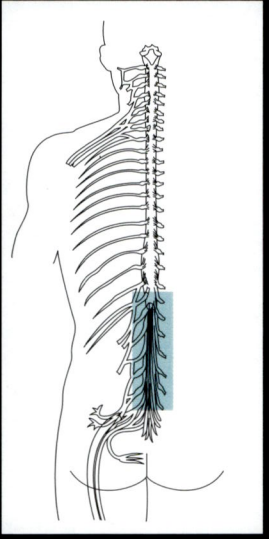

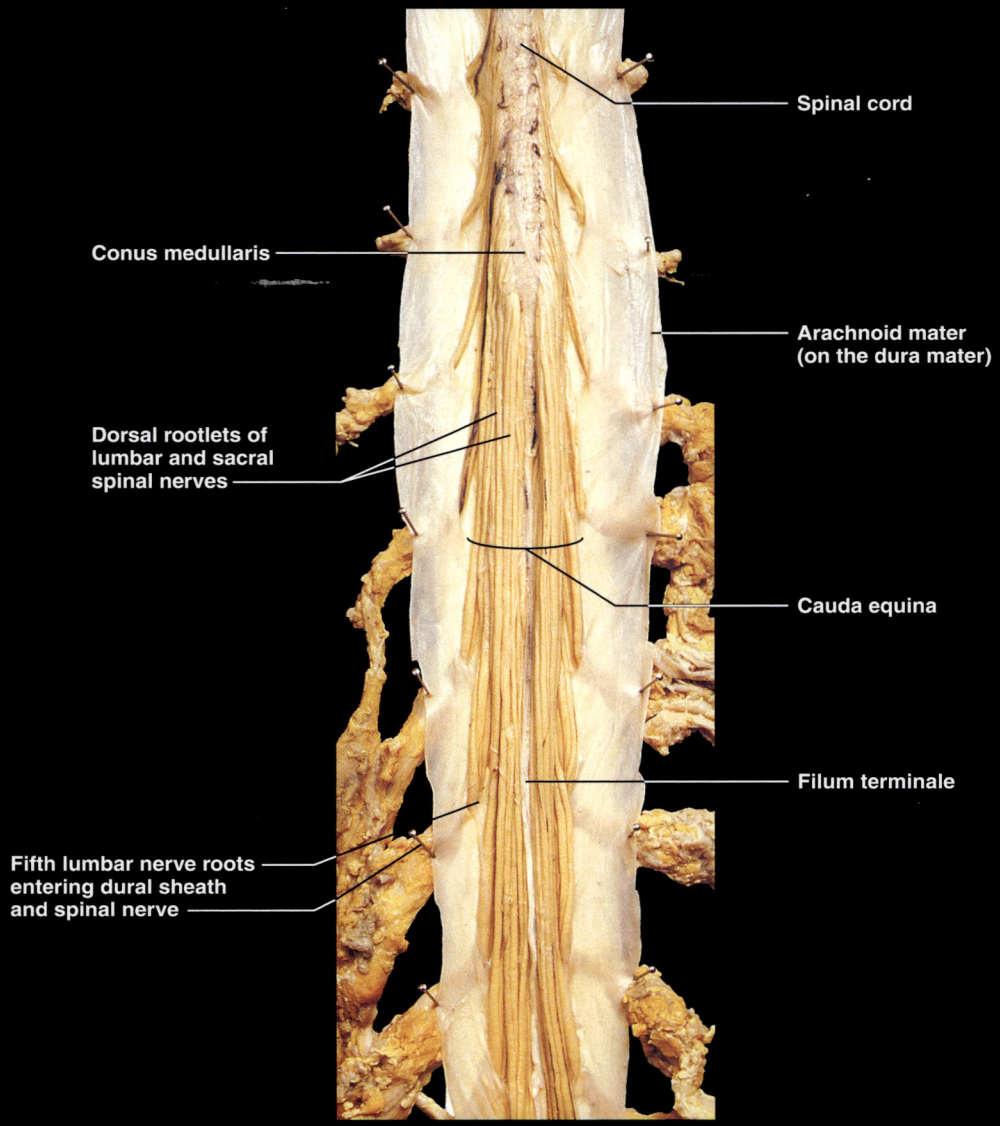

Figure 55 Spinal cord and cauda equina, dorsal view of lower end.

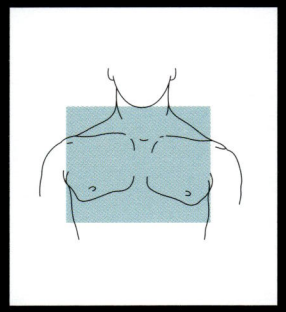

Superior vena cava
Right auricle
Right atrium
Right lung
Pericardium
Right ventricle
Diaphragm
Left brachiocephalic vein
Ascending aorta
Left lung
Pulmonary trunk
Pericardium
Left ventricle
Fat in epicardium
Apex of the heart

Figure 56 Heart and associated structures in thorax.

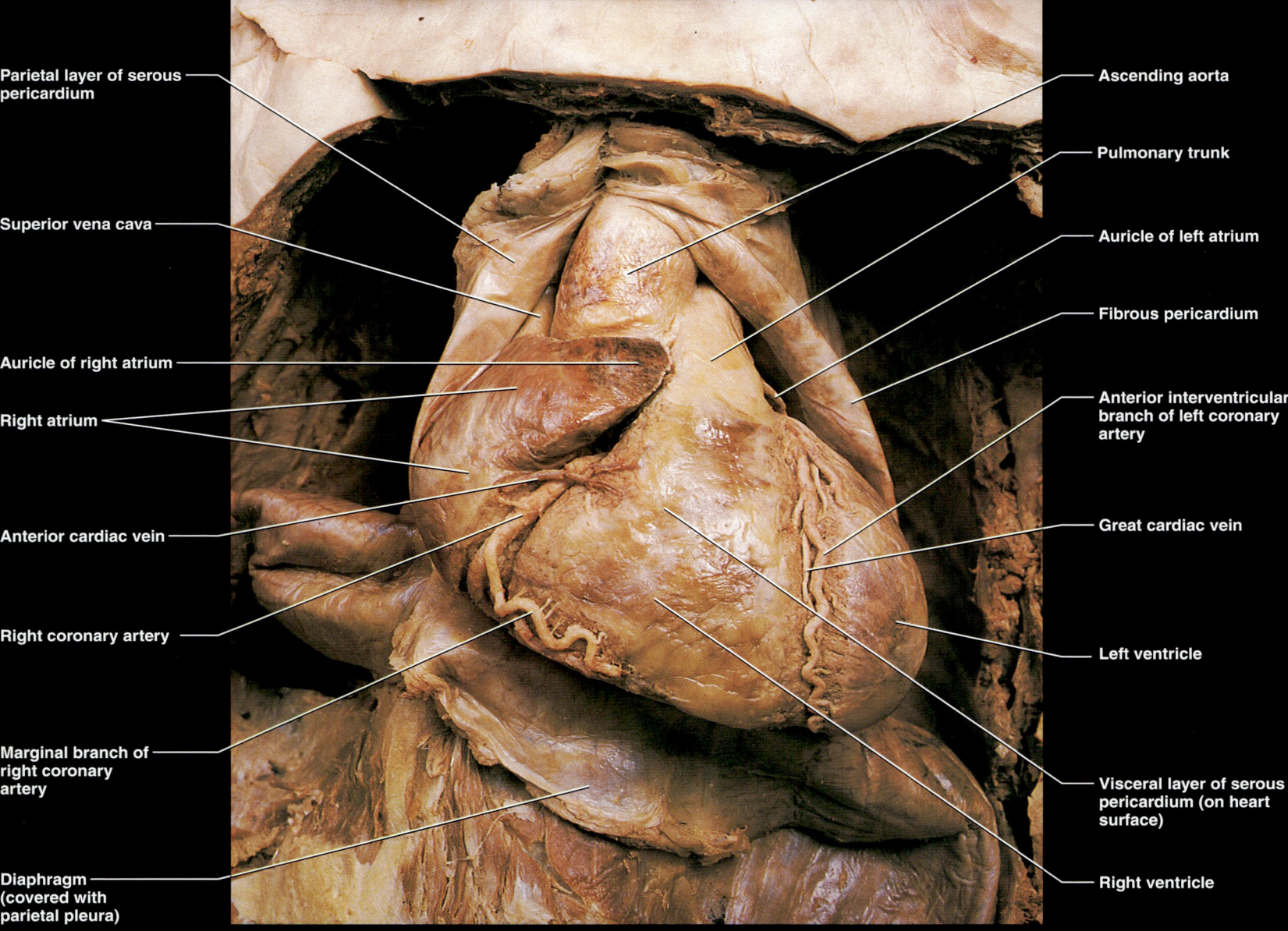

Figure 57 Heart and pericardium, anterior view.

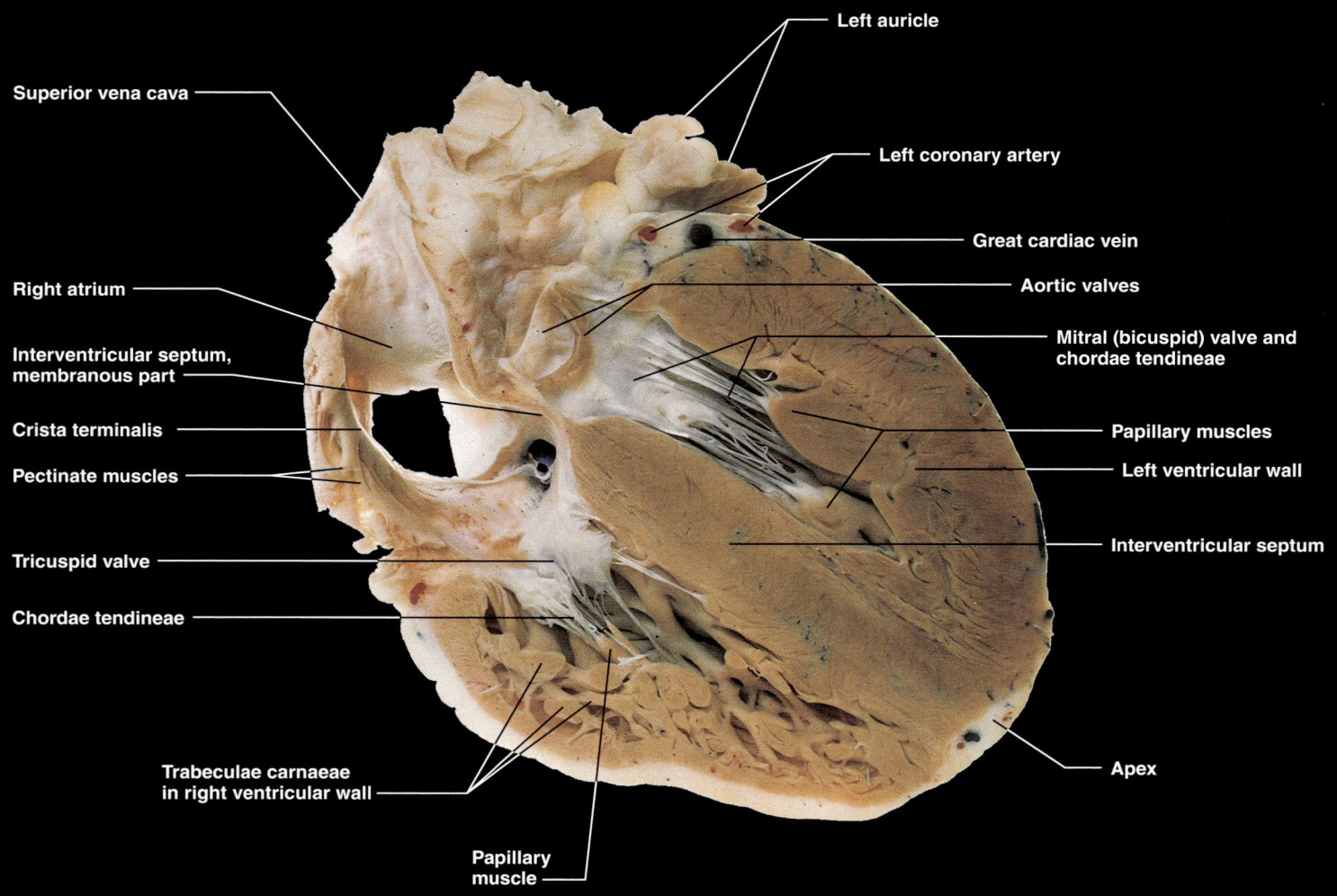

Figure 58 Coronal section of the ventricles, anterior view.

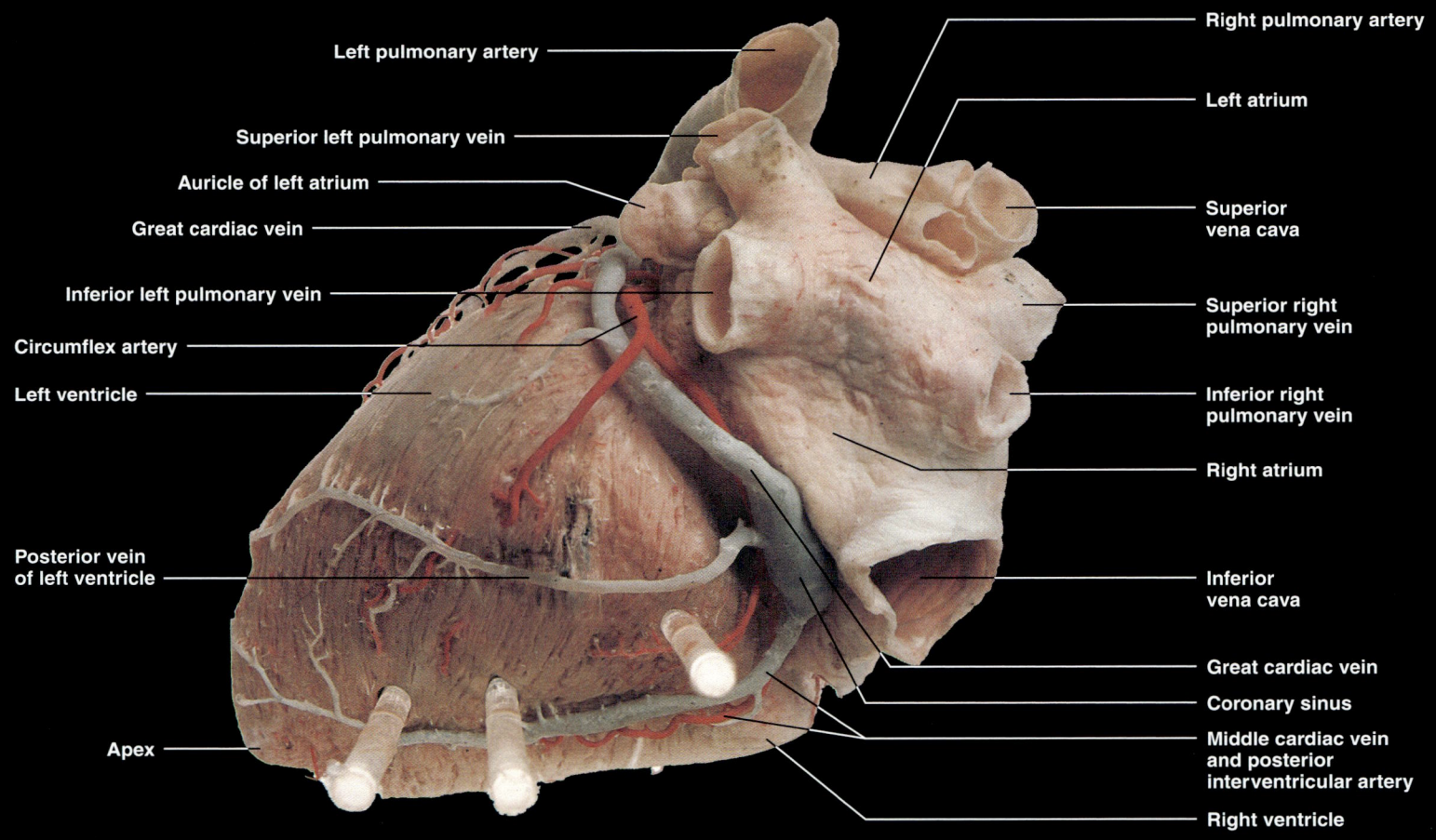

Figure 59 Heart, posterior view (blood vessels injected).

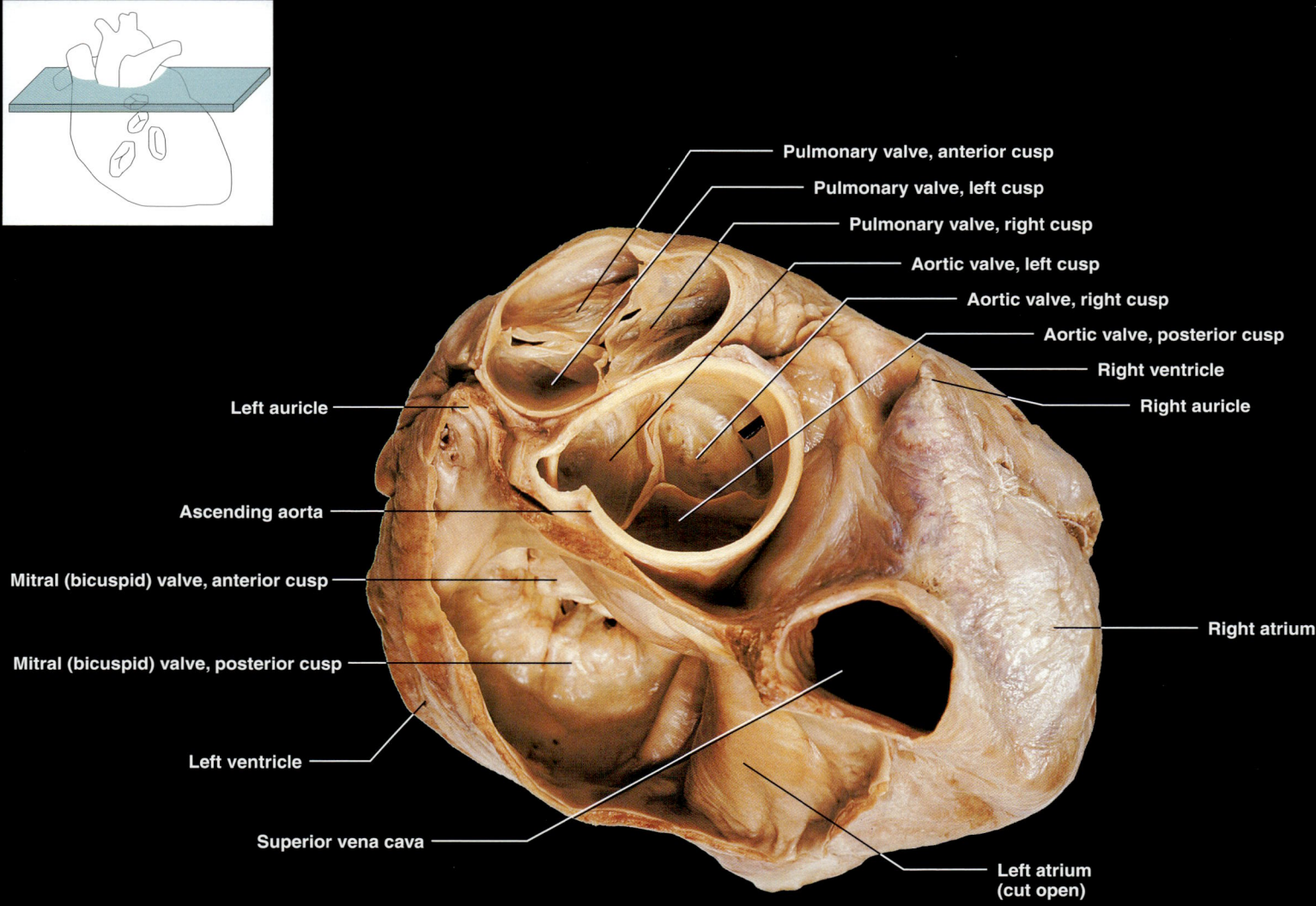

Figure 60 Pulmonary, aortic, and mitral valves of the heart, superior view.

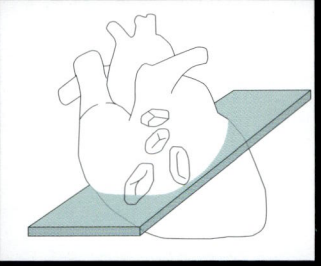

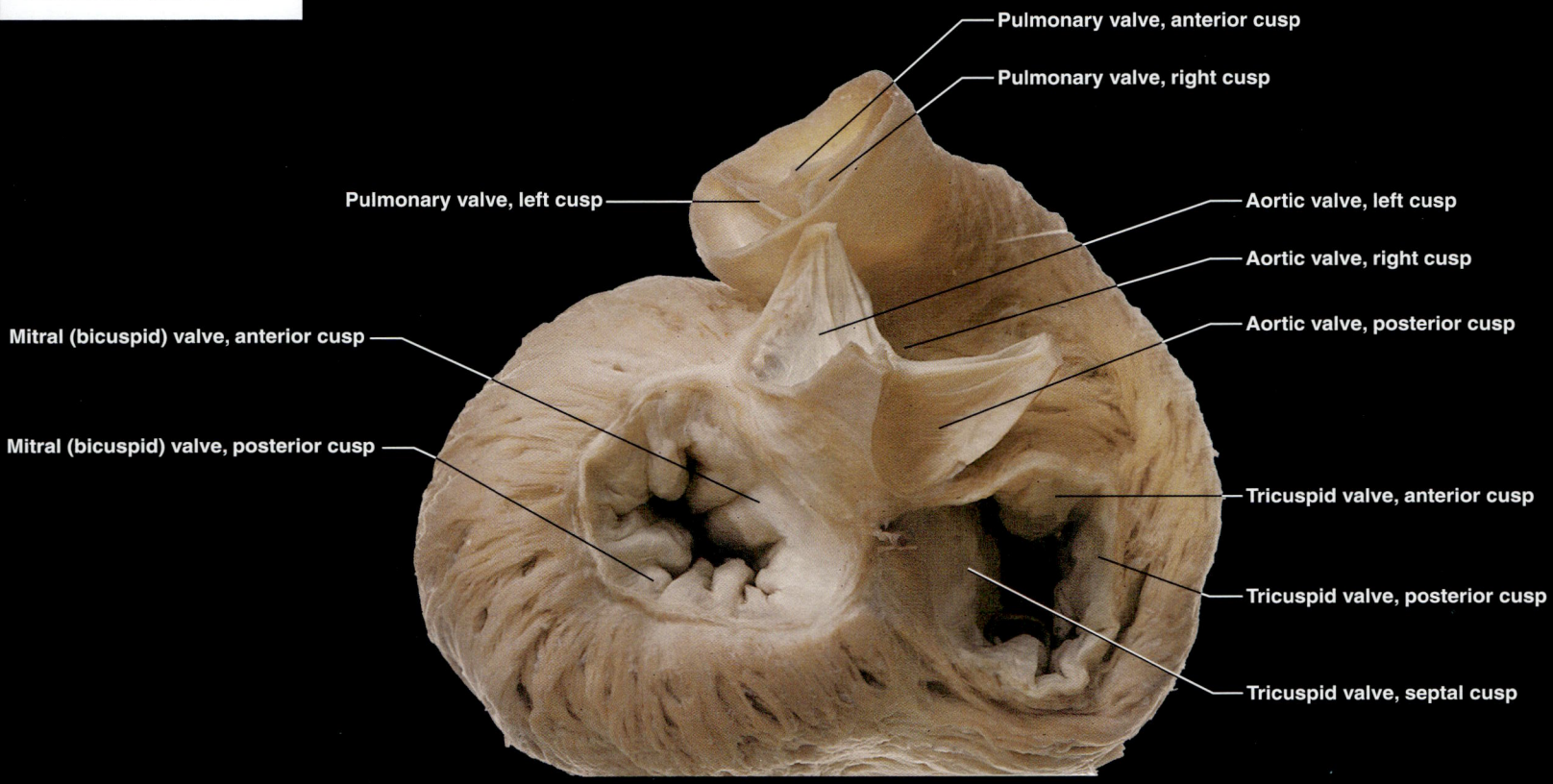

Figure 61 Fibrous framework of the heart (atria removed), posterior view, from the right.

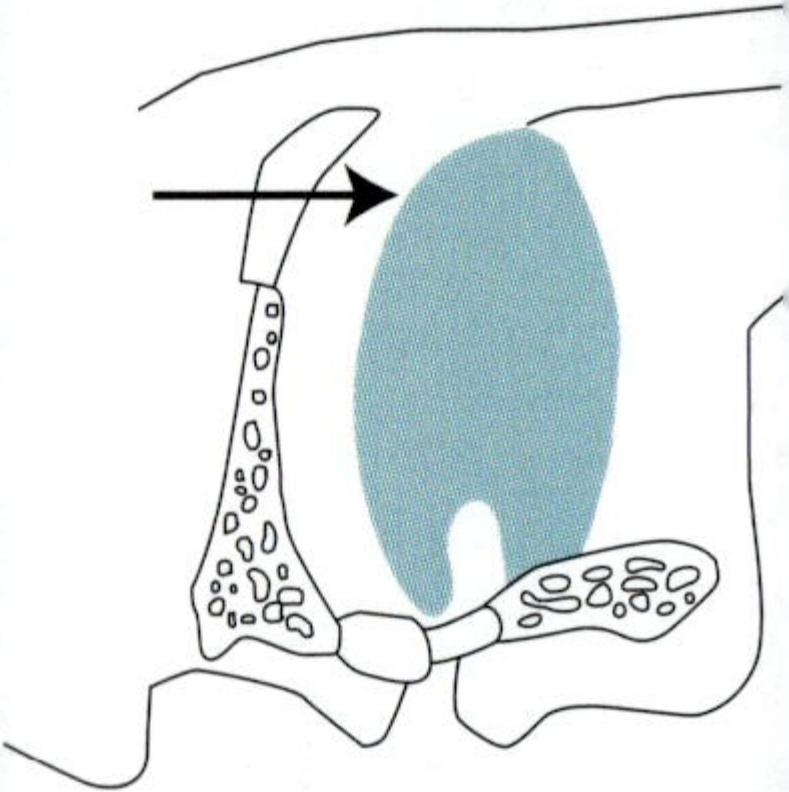

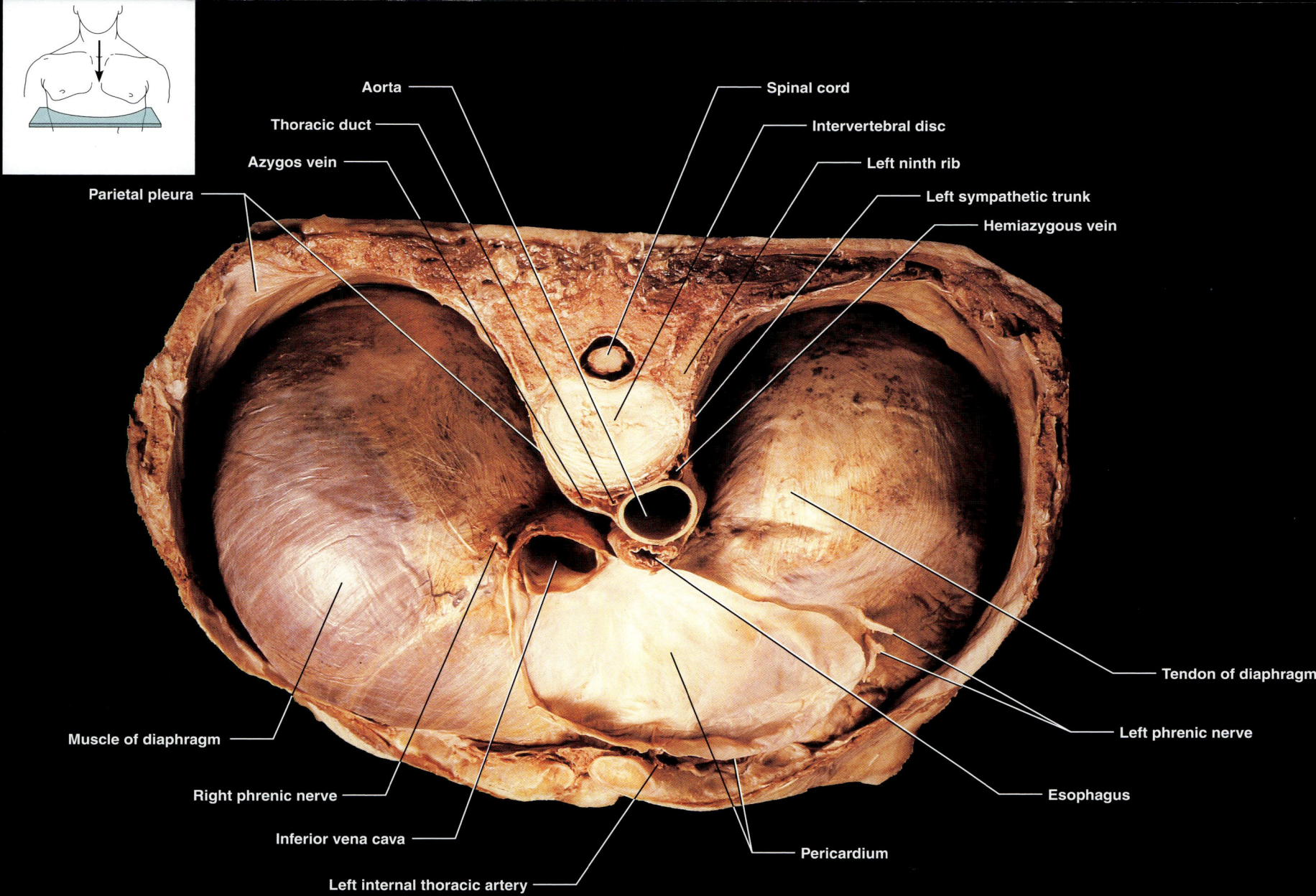
Aorta
Thoracic duct
Azygos vein
Parietal pleura
Spinal cord
Intervertebral disc
Left ninth rib
Left sympathetic trunk
Hemiazygous vein
Tendon of diaphragm
Left phrenic nerve
Esophagus
Pericardium
Muscle of diaphragm
Right phrenic nerve
Inferior vena cava
Left internal thoracic artery

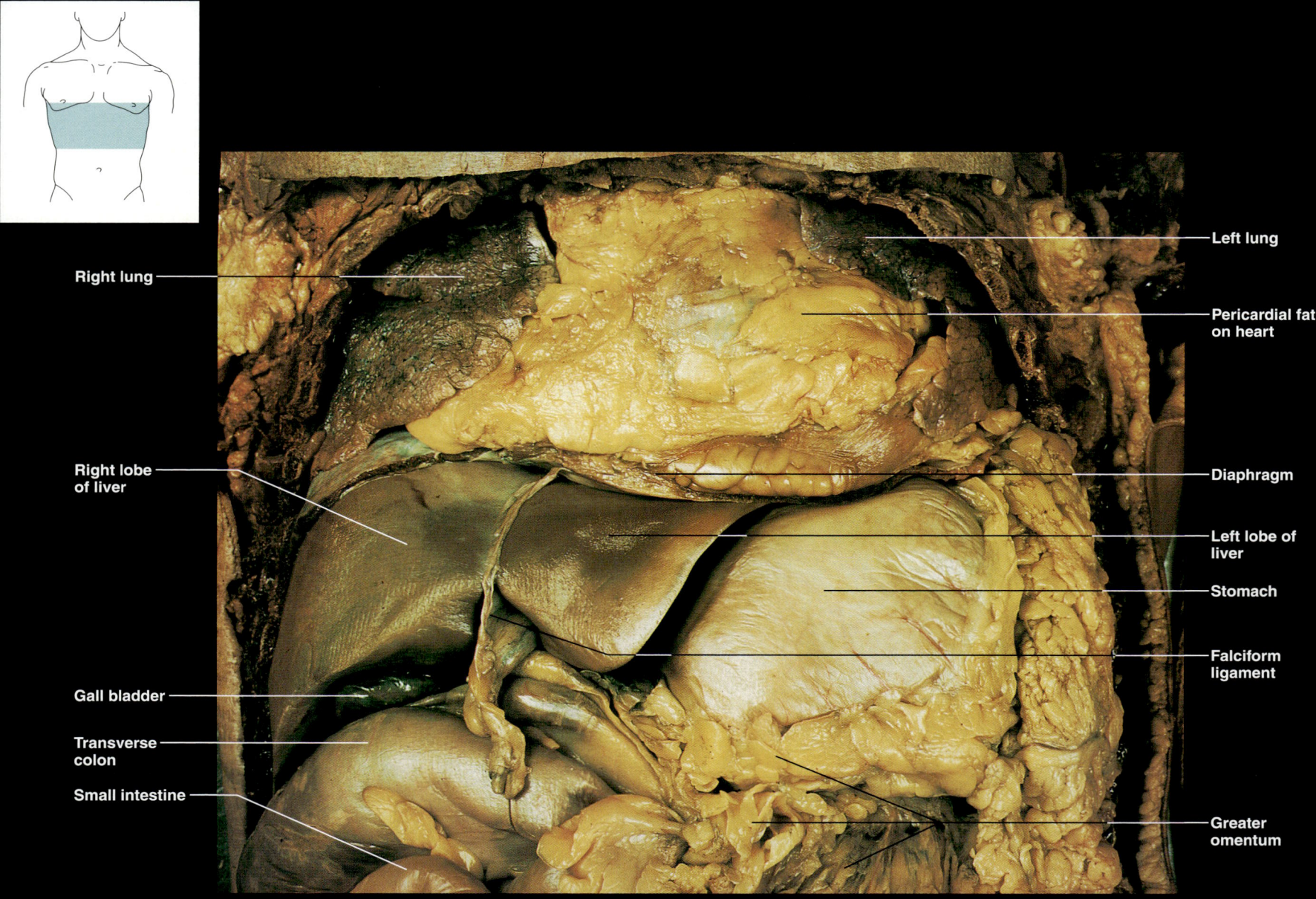

(a) upper abdominal viscera, anterior view

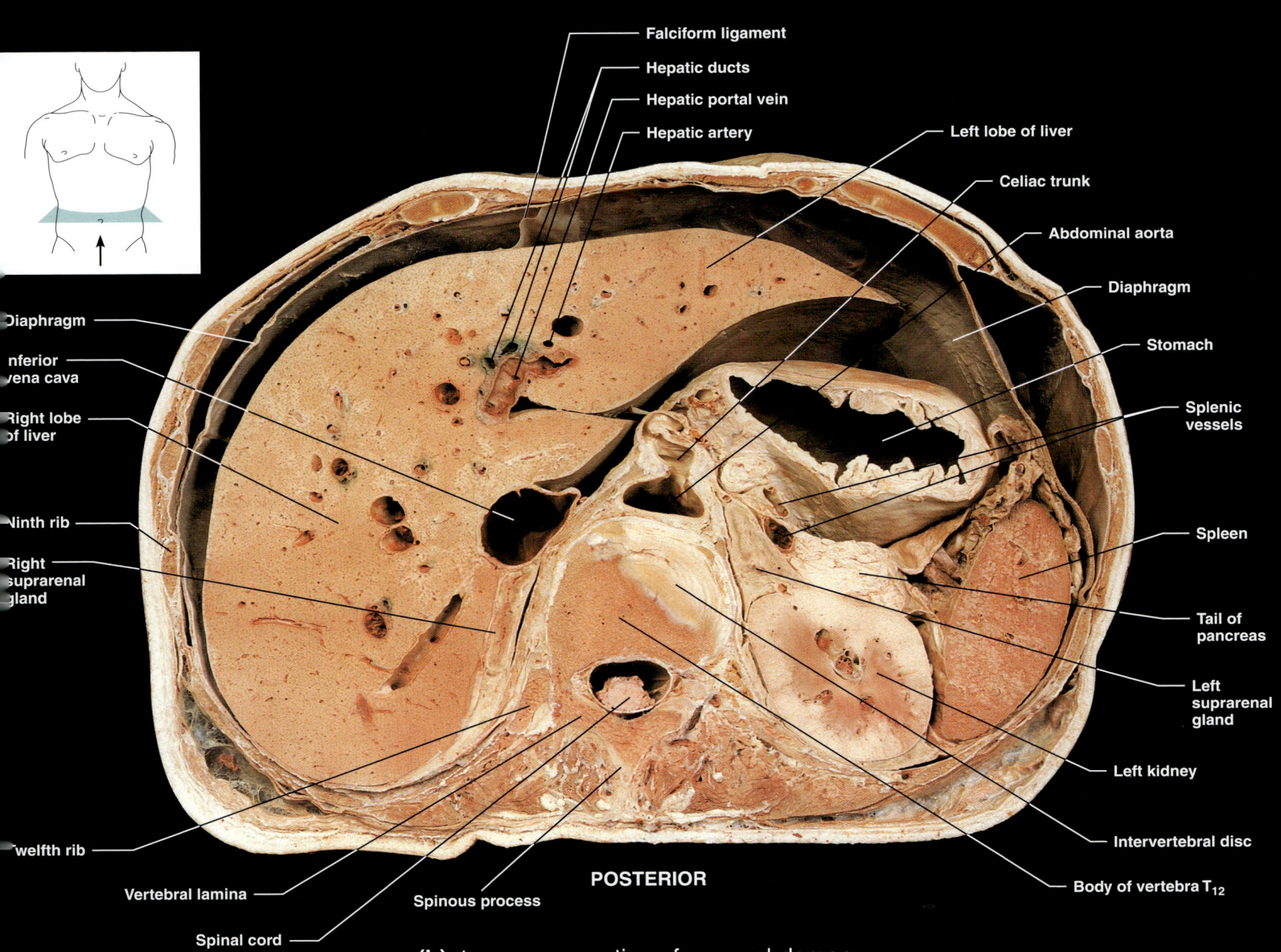

(b) transverse section of upper abdomen, from below, showing liver, stomach, and spleen *in situ*

Figure 64 Upper abdomen.

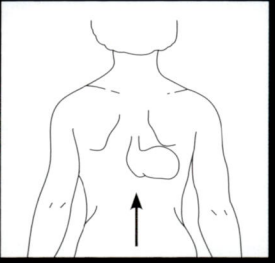

Left triangular ligament

Diaphragm

Inferior vena cava

Bare area

Caudate lobe

Porta hepatis

Right lobe

Ligamentum venosum in fissure

Left lobe

Hepatic artery

Hepatic portal vein

Common hepatic duct

Ligamentum teres in the falciform ligament

Quadrate lobe

Gallbladder

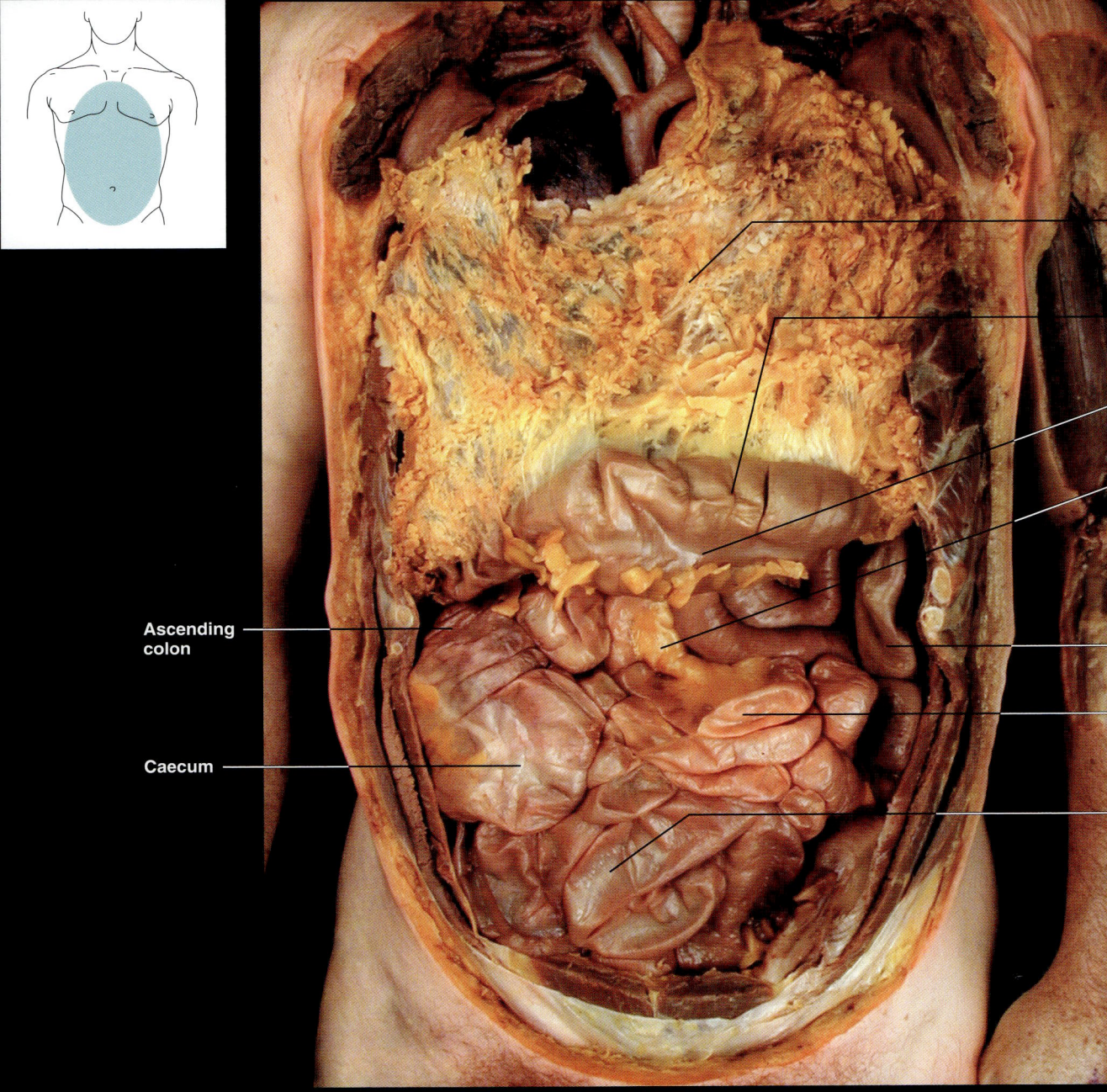
Ascending colon
Caecum

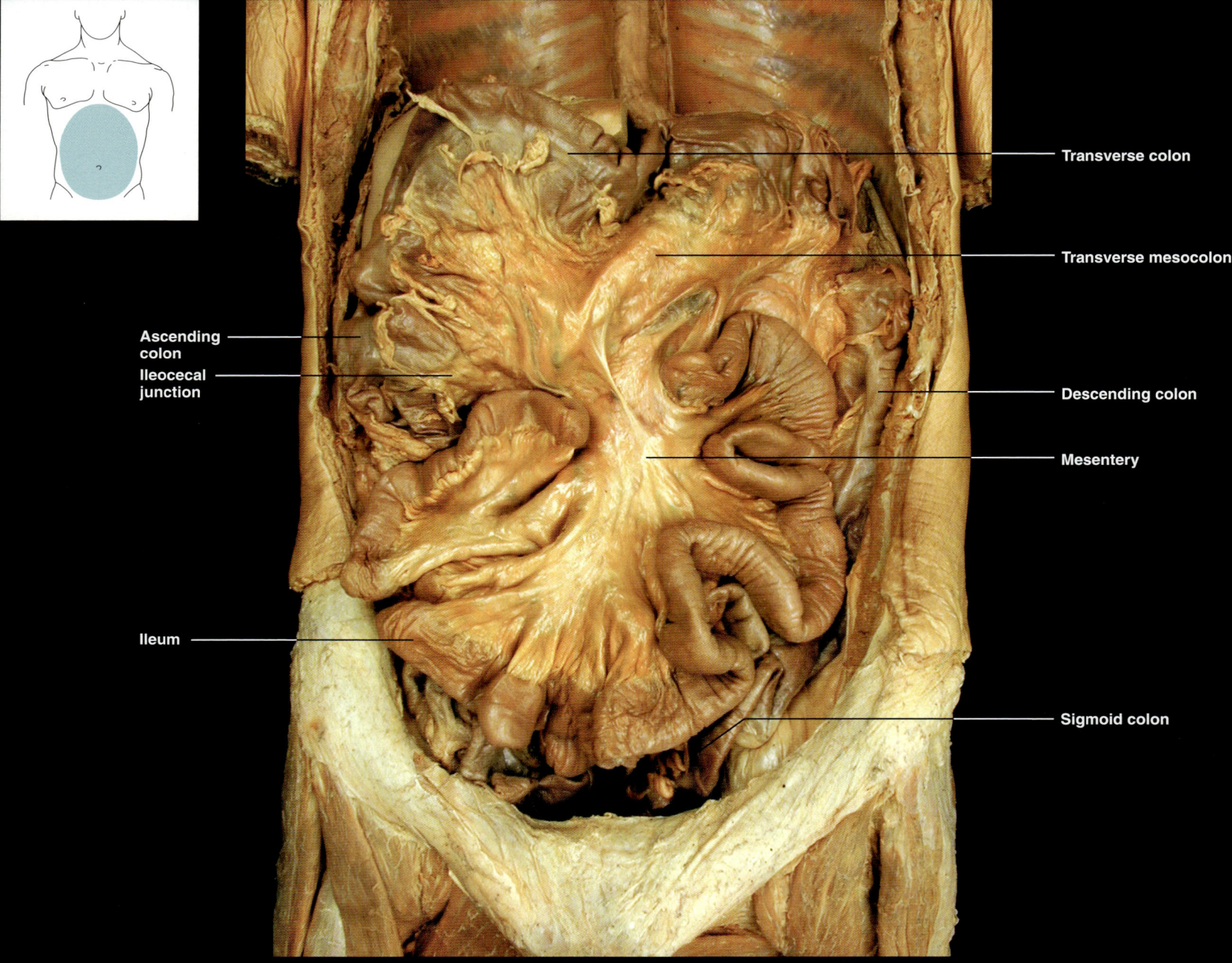

Figure 67 Small intestine and colon.

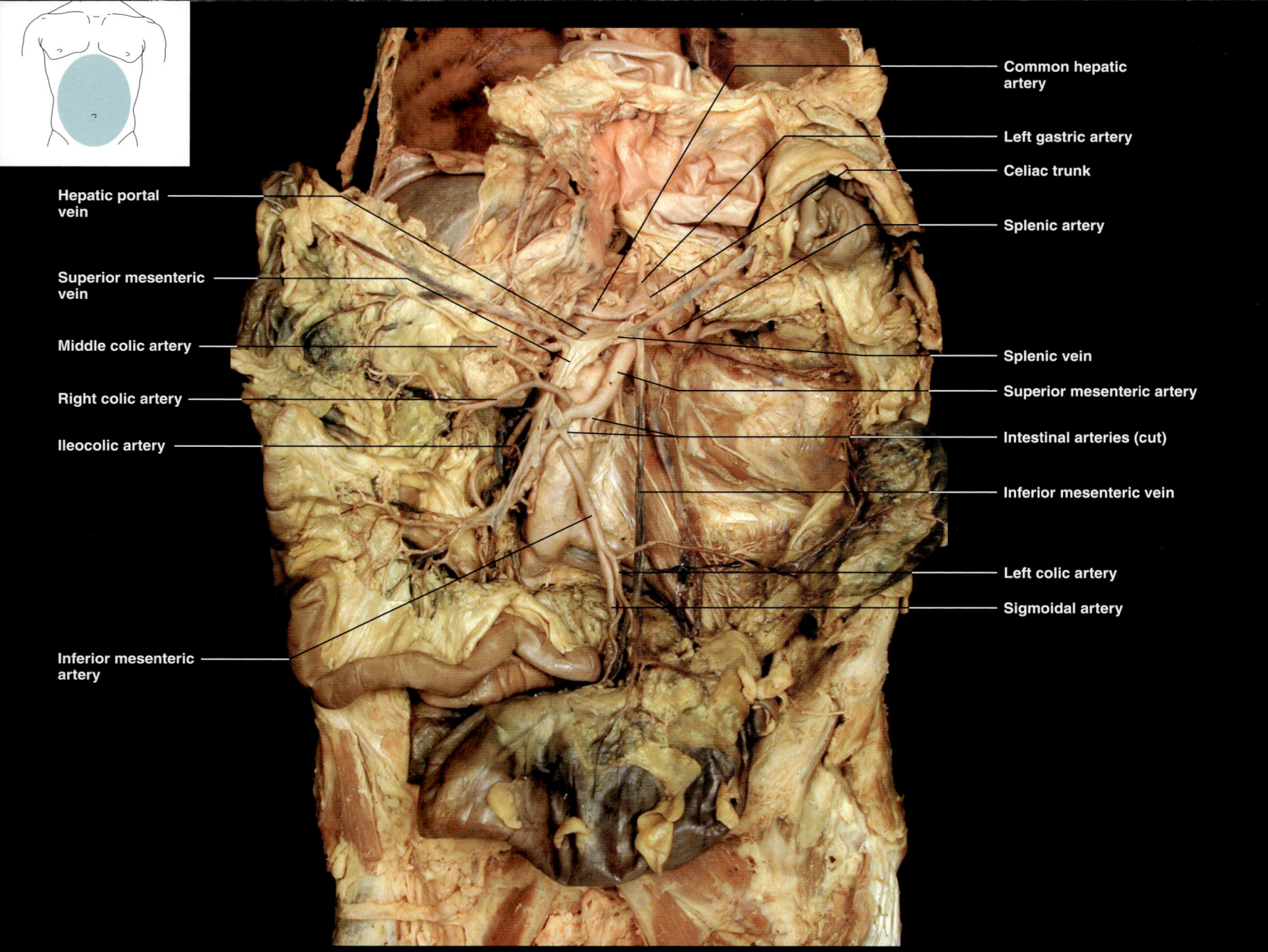

Figure 68 Vessels of the gastrointestinal organs.

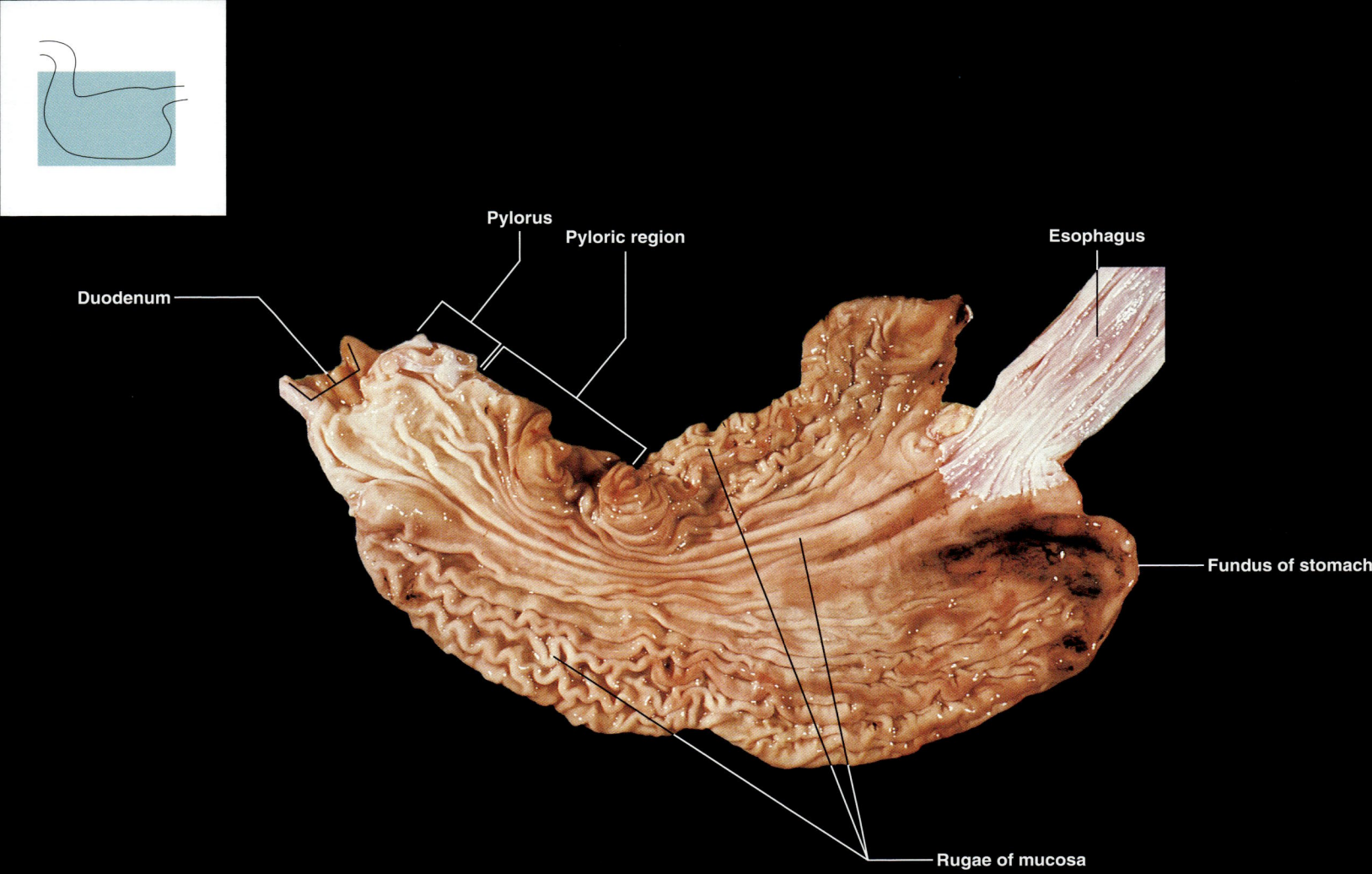

(a) frontal section of the internal surface of the stomach.

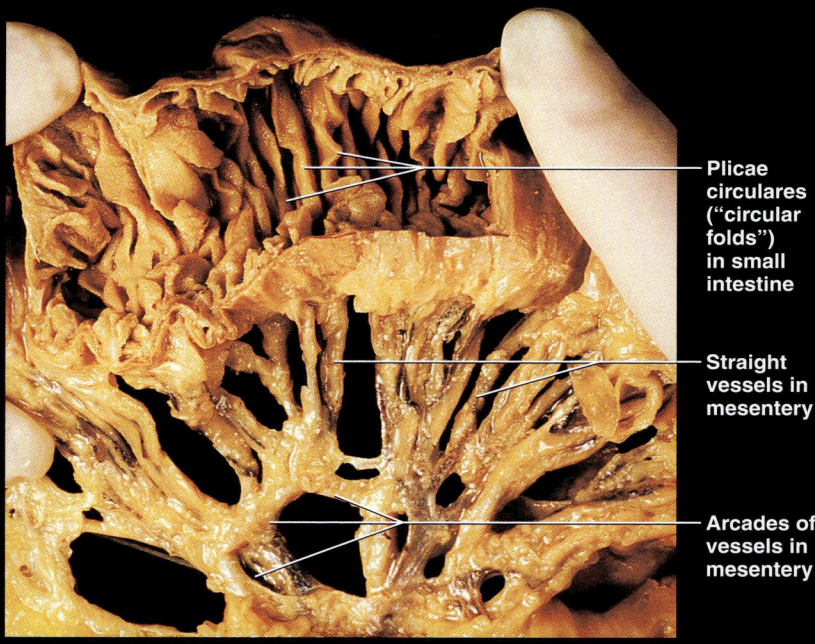

(b) small intestine, cut open to show plicae circulares

Figure 69 Internal surfaces of the stomach and small intestine.

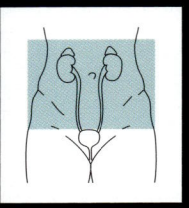

Diaphragm
Hepatic vein
Inferior vena cava
Renal veins
Right testicular artery
Inferior mesenteric artery
Common iliac arteries
Common iliac veins
Right external iliac artery
Right internal iliac artery
Abdominal aorta
Celiac trunk
Superior mesenteric artery
Suprarenal (adrenal) gland
Left middle suprarenal artery
Kidney
Left renal arteries
Quadratus lumboreum
Psoas minor
Psoas major
Ureter
Urinary bladder

Figure 70 Retroperitoneal abdominal structures.

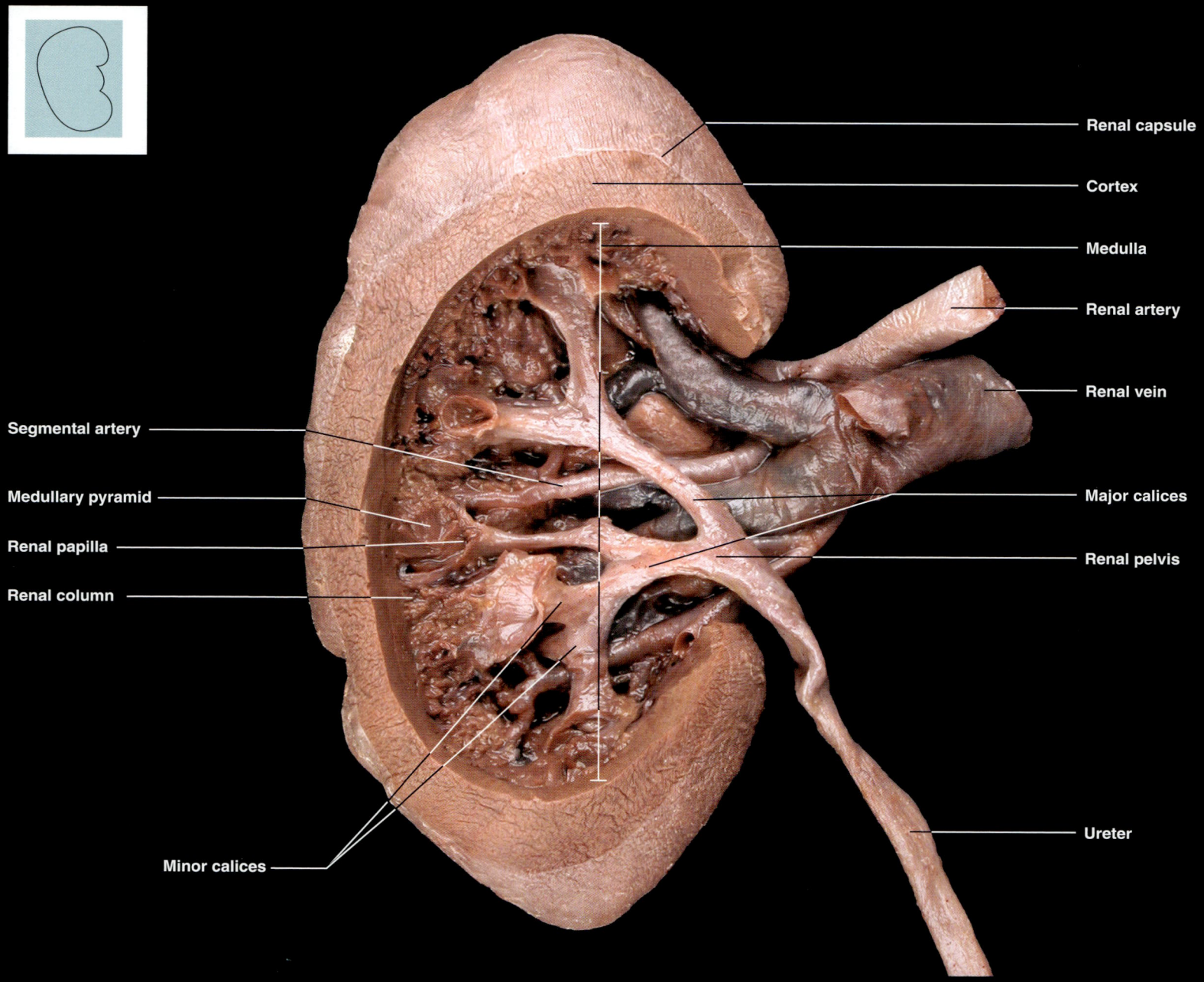

Figure 71 Kidney, internal structure in frontal section

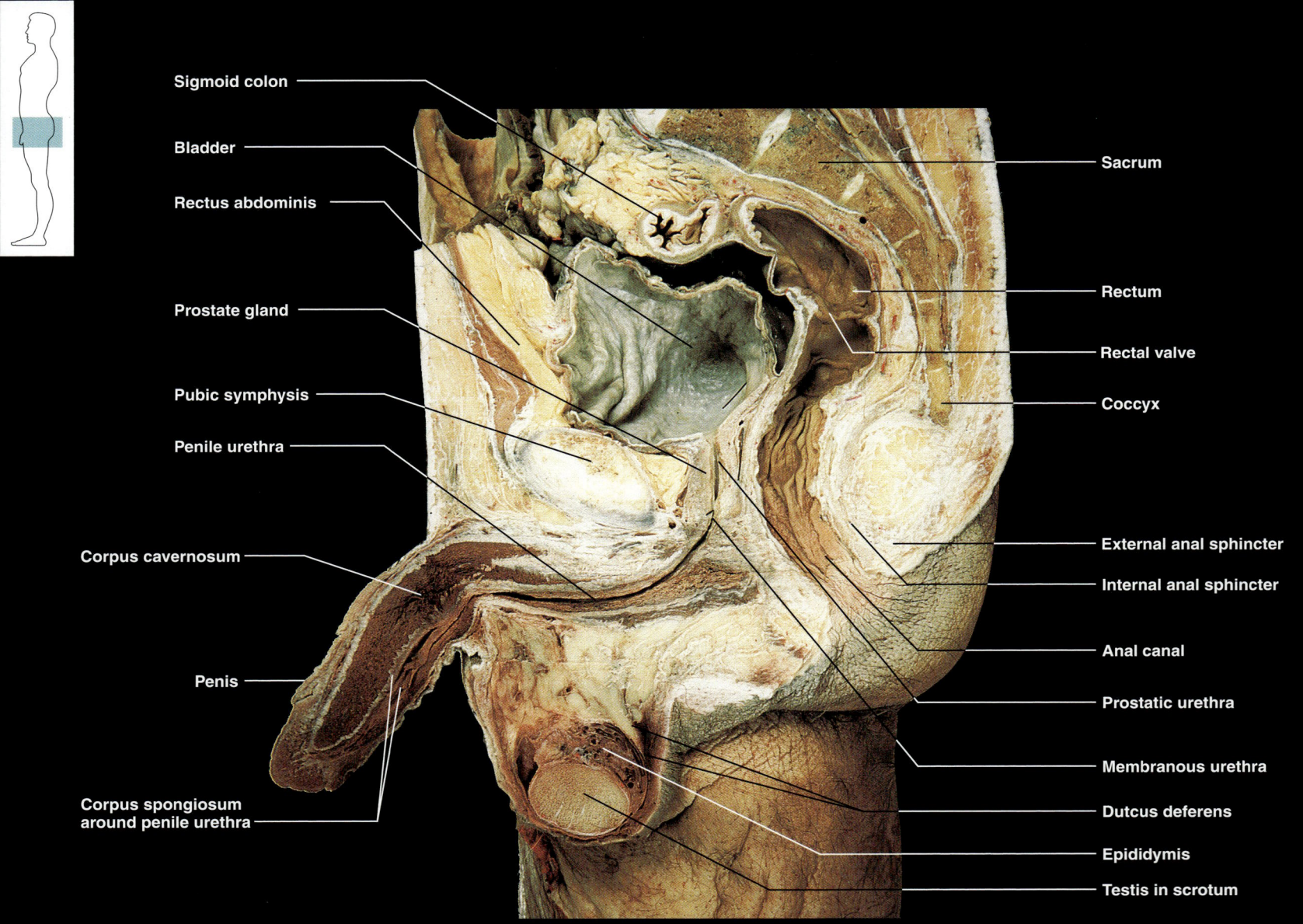

Figure 72 Male pelvis, sagittal section.

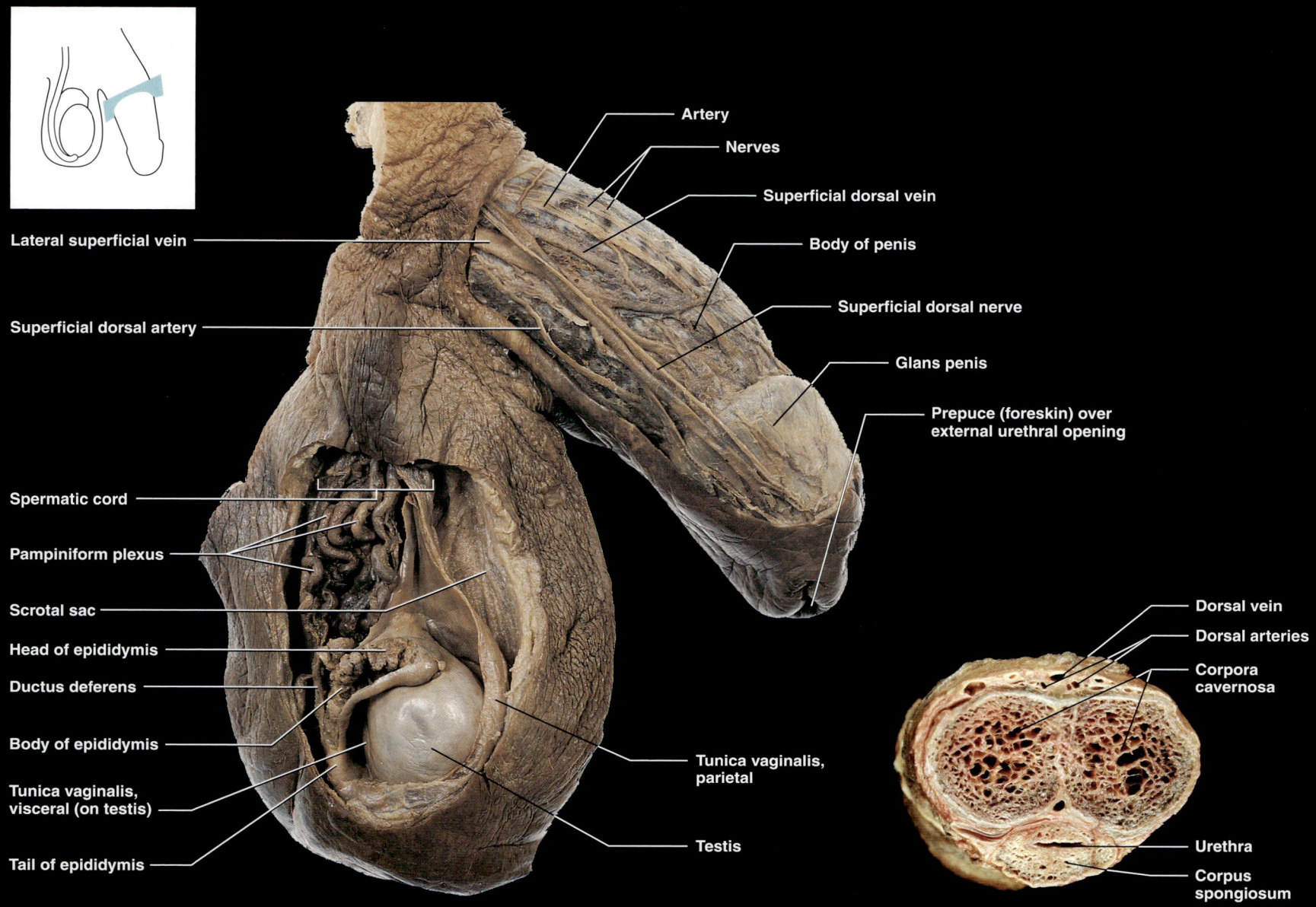

(a) section through right testis and epididymis, and the penis

(b) transverse section through penis

Figure 73 Sections through male reproductive structures.

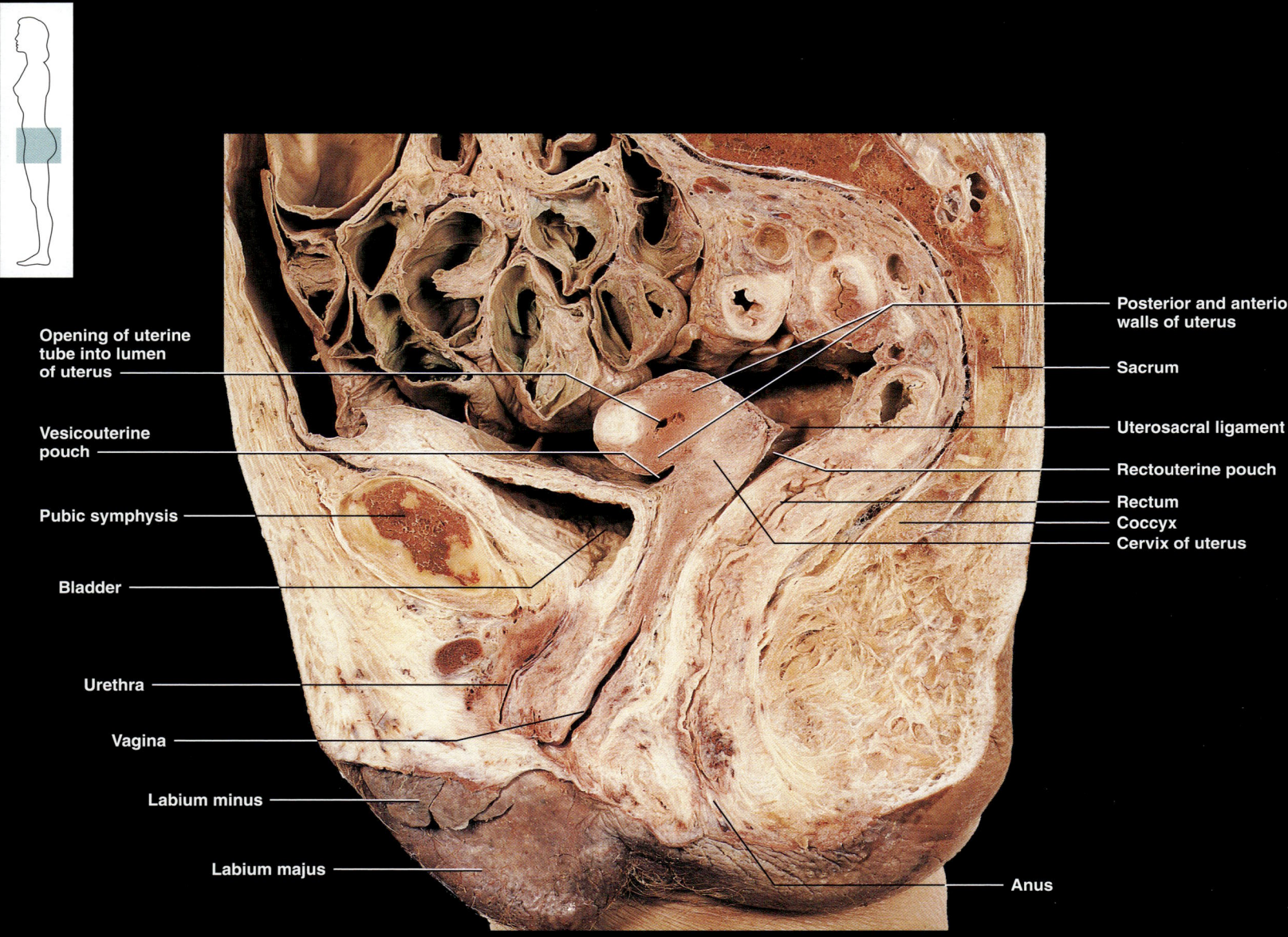

Figure 74 Female pelvis, sagittal section (uterus points forward in this view).

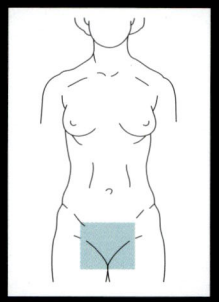

Recto-uterine pouch

Fundus of uterus

Mesometrium of broad ligaments, posterior surface

Tubal extremity of right ovary

Uterine extremity of right ovary

Ovarian ligament

Rectum

Left ovary

Infundibulum of uterine tube

Suspensory ligament of ovary

Mesovarium

Ampulla of uterine tube

Mesosalpinx

Isthmus of uterine tube

Round ligament

Ovarian ligament

Vesicouterine pouch

Bladder

Figure 75 Female pelvic cavity showing the position of the uterus relative to other structures.